Studies in Economic History

Aims and Scope

This series from Springer provides a platform for works in economic history that truly integrate economics and history. Books on a wide range of related topics are welcomed and encouraged, including those in macro-economic history, financial history, labor history, industrial history, agricultural history, the history of institutions and organizations, spatial economic history, law and economic history, political economic history, historical demography, and environmental history.

Economic history studies have greatly developed over the past several decades through application of economics and econometrics. Particularly in recent years, a variety of new economic theories and sophisticated econometric techniques—including game theory, spatial economics, and generalized method of moment (GMM)—have been introduced for the great benefit of economic historians and the research community.

At the same time, a good economic history study should contribute more than just an application of economics and econometrics to past data. It raises novel research questions, proposes a new view of history, and/or provides rich documentation. This series is intended to integrate data analysis, close examination of archival works, and application of theoretical frameworks to offer new insights and even provide opportunities to rethink theories.

The purview of this new Springer series is truly global, encompassing all nations and areas of the world as well as all eras from ancient times to the present. The editorial board, who are internationally renowned leaders among economic historians, carefully evaluate and judge each manuscript, referring to reports from expert reviewers. The series publishes contributions by university professors and others well established in the academic community, as well as work deemed to be of equivalent merit.

More information about this series at http://www.springer.com/series/13279

Wayne Geerling • Gary Magee

Quantifying Resistance

Political Crime and the People's Court in Nazi Germany

Springer

Wayne Geerling
Department of Economics
University of Arizona
Tucson, AZ, USA

Gary Magee
Department of Economics
Monash University
Caulfield East, VIC, Australia

ISSN 2364-1797 ISSN 2364-1800 (electronic)
Studies in Economic History
ISBN 978-981-10-6007-6 ISBN 978-981-10-6008-3 (eBook)
DOI 10.1007/978-981-10-6008-3

Library of Congress Control Number: 2017949270

Printed on acid-free paper

This Springer imprint is published by Springer Nature
The registered company is Springer Nature Singapore Pte Ltd.
The registered company address is: 152 Beach Road, #21-01/04 Gateway East, Singapore 189721, Singapore

Foreword

One of the exciting – if at times frustrating – things about doing research is its unpredictability. One can never know precisely what one will uncover until it is actually done. And there are always surprises and disappointments along the way. Indeed, as most seasoned researchers would no doubt attest, serendipity plays a significant part in the process of intellectual discovery. In an academic world now accustomed to talking in terms of the certainty of research strategies, management and planning, it may not be fashionable to say as much. But it is indubitably true.

The project that gave rise to this book has had its serendipitous moments, too. In fact, it originated in a completely unpremeditated manner out of a casual conversation between the authors about one of their PhD theses, which had been on juvenile resistance and opposition in Nazi Germany. From that discussion, two interesting observations unwittingly emerged.

First, it was noted that while there existed an extensive and high-quality literature on internal resistance in Nazi Germany, there were nonetheless few books that gave a comprehensive overview of the relative contributions made by different types of resisters to the overall story of serious resistance. Yes, there were a number of very good accounts which listed the various resistance groups operative in Nazi Germany in a careful, encyclopaedic manner, but few of these, we felt, allowed a reader to assess which of these varieties of resistance were more prominent and impactful in different periods of the regime's existence. We saw this as a significant gap in the literature and figured that by adopting a systematic, social scientific approach, such a concise account of resistance, which accurately placed all forms of resistance in an overarching story, might just be possible.

Second, we also realised that the source material used in the aforementioned PhD on juveniles – court and investigation records for treason and high treason cases – contained, albeit in an unprocessed form, information, which, if properly collated, extended and analysed, could provide a solid quantitative foundation for such an account of serious resistance. We resolved to explore these insights further to see where it might take us.

This book is a fruit of those efforts. We hope that upon reading, you will see it as a useful addition to our knowledge and understanding of serious resistance in Nazi

Germany and believe that it has gone at least some of the way toward realising the goals envisaged at the project's inception. Time will tell.

No research is created in an intellectual vacuum. This project is no exception. We acknowledge that the Quantifying Resistance Project would not have progressed so far had it not been for the generous support, comments and advice of a host of friends and colleagues. We are deeply indebted to many, not least among whom are David Addison, Friederike Bauer, Chongwoo Choe, Rosalind Hearder, Kim Huynh, Sisira Jayasuriya, Steffen Jöris, Stephen King, Alice Li, Stuart Macintyre, Michael Smiddy, Hans-Joachim Voth, Steven Welch, Wolfgang Wippermann, Jürgen Zarusky and the librarians and archivists in Germany, the United States and Australia who are far too numerous to mention individually. Special thanks also go to those colleagues who have joined us along the way to write papers that have emerged, often largely unexpectedly, from the project and from whom we have learnt a great deal: Rob Brooks, Paul Raschky, Russell Smyth and Vinod Mishra. While their contributions and insights are many, the errors that remain are ours and ours alone.

For us, perhaps the most striking feature of our research into the resistance against Hitler has been the great diversity of people who, for one reason or another, were drawn into it. The attributes needed to stand up to the terror and brutality of the Nazi regime clearly did not belong solely to any one type of individual. Germans and Austrians of all backgrounds and persuasions defied the tyrannical. If, by writing this book, we are able to stir interest in others to remember, appreciate and seek out more knowledge about their bravery, then our research will have been worthwhile.

Contents

List of Abbreviations

AAM	*Antifaschistische Arbeitergruppe Mitteldeutschlands*
AFÖ	*Antifaschistische Freiheitsbewegung Österreichs*
BBC	British Broadcasting Corporation
BDM	*Bund Deutscher Mädel*
CCL	Conservative, Catholic and Legitimist groups
CSP	*Christlichsoziale Partei*
DFP	*Deutsche Freiheitspartei*
DMV	*Deutscher Metallarbeiter-Verband*
DÖW	*Dokumentationsarchiv des österreichischen Widerstandes*
FAUD	*Freie Arbeiter Union Deutschlands*
FRG	Federal Republic of Germany
GDR	German Democratic Republic
GÖFB	*Großösterreichische Freiheitsbewegung*
HJ	*Hitlerjugend*
IÖKF	*Illegale österreichische kaiserliche Front*
IKD	*Internationale Kommunisten Deutschlands*
ISK	*Internationaler Sozialistischer Kampfbund*
KJVD	*Kommunistische Jugendverband Deutschlands*
KJVÖ	*Kommunistische Jugendverband Österreichs*
KPD	*Kommunistische Partei Deutschlands*
KPDO	*Kommunistische Partei-Opposition*
KPÖ	*Kommunistische Partei Österreichs*
NKFD	*Nationalkomitee freies Deutschland*
NKVD	People's Commissariat of Internal Affairs
NSDAP	*Nationalsozialistische Deutsche Arbeiterpartei*
ÖFB	*Österreichische Freiheitsbewegung*
ÖFF	*Österreichische Freiheitsfront*
OLG	*Oberlandesgericht*
RAF	*Rote Armee Fraktion*
RAF	Royal Air Force
RGBl	*Reichsgesetzblatt*

RFB	*Roter Frontkämpferbund*
RGO	*Revolutionäre Gewerkschafts-Opposition*
RSÖ	*Revolutionäre Sozialisten Österreichs*
SA	*Sturmabteilung*
SAJ	*Sozialistische Arbeiterjugend*
SAP	*Sozialistische Arbeiterpartei Deutschlands*
SPD	*Sozialdemokratische Partei Deutschlands*
SPÖ	*Sozialdemokratische Partei Österreichs*
SS	*Schutzstaffel*
USAAF	United States Army Air Force
USSR	Union of Soviet Socialist Republics
VGH	*Volksgerichtshof*
WKZ	*Wehrkraftzersetzung*

Glossary of German Terms

Alltagsgeschichte social history from below

***Alter Kämpfer* (singular)/*Alte Kämpfer* (plural)** old Nazi party fighters, who joined the party before the Nazis came to power

Altreich term used after the annexation of Austria in 1938 to refer to German territory within the 1937 (pre-annexation boundaries)

Amtsgerichtsrat local court judge

angelernter Arbeiter trained worker

Anklage indictment

Anschluß Germany's unification with (annexation of) Austria in March 1938

***Antifaschistische Arbeitergruppe Mitteldeutschlands* (*AAM*)** Anti-Fascist Workers Group of Central Germany, a communist-organised resistance group

***Antifaschistische Freiheitsbewegung Österreichs* (*AFÖ*)** Anti-Fascist Freedom Movement for Austria, a national-conservative resistance group

Bündische Jugend a movement of many different youth associations, including hiking groups and boy scouts, which arose after WWI

***Bund Deutscher Mädel* (*BDM*)** League of German Girls, the female branch of the Hitler Youth

Bundesarchiv Federal Archive

Bundesländer federal states

Bundesrepublik Federal Republic of West Germany

Bürgerschule lower form of secondary education

***Christlichsoziale Partei* (*CSP*)** Christian Social Party, a conservative party in Austria, which operated from 1891 to 1934

Christlich-Soziale Reichspartei Christian Socialist Reich Party, a left-wing Catholic party active during the Weimar Republic

Datenschutz protection of privacy law

***Deutsche Freiheitspartei* (*DFP*)** German Freedom Party, a national conservative resistance group founded in exile in 1937

***Deutscher Metallarbeiter-Verband* (*DMV*)** German Metal Workers Trade Union

Dokumentationsarchiv des österreichischen Widerstandes (DÖW) Documentation Centre of Austrian Resistance

Einheitsfront Unity Front, a strategy of the Comintern where the ***KPD*** would co-operate with other left-wing groups as part of a broader anti-fascist front

Erschliessungsband index book

Europäische Union European Union, an anti-fascist resistance group founded in Berlin in 1939

Feindbegünstigung giving aid and support to the enemy

Freie Arbeiter Union Deutschlands (FAUD) Free Workers Union of Germany, which became an anarcho-syndicalist resistance group in the 1930s

Freikorps Free Corps, paramilitary organisations that sprang up around Germany as soldiers returned in defeat from WWI

Freies Sozialistische Jugend Free Socialist Youth, a political youth organisation which existed from 1918 to 1920

Führer leader

Gedenkstätte Deutscher Widerstand German Resistance Memorial Center

Gefängnis prison

Generaloberst colonel general

Geselle journeyman, a term used to describe a trained worker who has finished an apprenticeship but is not yet a master craftsman

Gestapo Secret State Police

Großösterreichische Freiheitsbewegung (GÖFB) Great Austrian Freedom Movement, a resistance group whose members came from catholic-conservative backgrounds

Gymnasium high school which leads to an ***Abitur*** (final exam) and is the pathway to university

Gebietsleiter regional leader, a term used within communist underground cells

Hauptschule lower form of secondary education

Heimwehr Home Guard, a paramilitary guard which operated in Austria in the 1920s to 1930s

Hitlerjugend (HJ) Hitler Youth, the German youth organisation founded by the Nazi Party

Hochschule technical college

Illegale österreichische kaiserliche Front (IÖKF) Austrian Underground Monarchist Front, a legitimist resistance group which sought the restoration of the Hapsburg monarchy

Internationale Kommunisten Deutschlands (IKD) International Communists of Germany, a Trotskyist resistance group

Internationaler Sozialistischer Kampfbund (ISK) International Socialist Militant League, a socialist resistance group which splintered from the ***SPD***

Kaiserreich German state which existed between 1871 and 1918

Kinderfreunde Social Democratic youth organisation

Kommunistische Jugendverband Deutschlands (KJVD) Communist Youth Association of Germany

Kommunistische Jugendverband Österreichs (KJVÖ) Communist Youth Association of Austria
Kommunistische Partei Deutschlands (KPD) Communist Party of Germany
Kommunistische Partei Österreichs (KPÖ) Communist Party of Austria
Kommunistische Partei-Opposition (KPDO) Communist Party of Germany (opposition)
Lächerlich laughable
Landesgerichtsdirektor district court presiding judge
Landesgerichtsrat district court judge
Lebensraum living space
Lehre* or *Lehrzeit apprenticeship
Lehrling apprentice
Machtergreifung Nazi seizure of power
Maschinenfabrik Augsburg-Nürnberg (MAN) Augsburg-Nuremberg Machinery Factory
Meister master (craftsman)
Meuten youth gangs which first arose in Leipzig in 1938
Mischlinge someone with partial Jewish ancestry
Mitgliedskarten membership records
Mittelschule lower form of secondary education
Nacht und Nebel Night and Fog, a code for some prisoners that were to be disposed of extrajudicially, leaving no traces
Nationalkomitee freies Deutschland (NKFD) National Committee for a Free Germany, a German anti-Nazi organisation that operated in the Soviet Union during World War II
Nationalsozialistische Deutsche Arbeiterpartei (NSDAP) Nazi Party
Neu Beginnen New Beginning, an anti-fascist group formed in 1929 by members of the ***SPD***
Neues freies Österreich New Free Austria, a resistance group which arose in 1944 and drew its members from across party lines
NKVD Soviet Secret Police
Oberlandesgericht Higher Regional Court
Oberstaatsanwalt Director of Public Prosecution
Österreichische Freiheitsbewegung (ÖFB) Austrian Freedom Movement, a national conservative resistance group
Österreichische Freiheitsfront (ÖFF) Austrian Freedom Front, an anti-fascist resistance group formed by Austrian and German communist refugees in exile
Ostmärkische Sturmscharen a right-wing paramilitary group in Austria, founded in 1930, recruiting mainly from catholic youth
Personalakten personal files
Pflichtschule compulsory primary school education
Pflichtverteidiger a court-assigned defence counsel
Praktikum internship
Realschule lower form of secondary education

Reichsdeutsche German citizens of the Reich

Reichsgericht Supreme Court. The Supreme Court heard cases of high treason during the Weimar Republic and continued to do so in the Third Reich before the People's Court opened in August 1934

Reichsjustizministerium Reich Ministry of Justice

Reichskriegsgericht Supreme Military Court

Reichssicherheitshauptamt Reich Main Security Office

Reichsstrafgesetzbuch Federal Penal Code

Reichstag German parliament

Republikanischer Schutzbund Republican Protection League, an Austrian paramilitary organisation created by the Social Democratic Party in 1923

Revolutionäre Gewerkschafts Opposition (***RGO***) Revolutionary Trade Union Opposition, a free trade union with links to the ***KPD***, parts of which engaged in resistance against the Nazis

Revolutionäre Sozialisten Revolutionary Socialists, an anti-fascist resistance group formed within the ***SPD*** in 1933

Revolutionäre Sozialisten Österreichs (***RSÖ***) Revolutionary Socialists of Austria, an anti-fascist resistance group formed in Austria in 1934

Richterbriefe judges' letters circulated by the Ministry of Justice from October 1942

Rote Armee Fraktion Red Army Faction, a left-wing terrorist group founded in West Germany in 1970

Rote Hilfe Red Help, a communist association designed to support the families of political prisoners

Rote Jungfront Red Youth Front, a youth organisation of the Red Front Fighters' League

Roter Frontkämpferbund (***RFB***) Red Front Fighters' League, a paramilitary organisation under the leadership of the ***KPD***

Schutzstaffel (***SS***) Protection Squad, a Nazi paramilitary organisation

Schwarze Front Black Front, a right-wing political group formed by Otto Strasser after his expulsion from the Nazi Party in 1930

Senatspräsident Senate President

Sondergericht Special Court, set up by the Nazi Party in March 1933 to help combat the activities of opponents of the regime. These courts existed outside constitutional law

Sozialdemokratische Partei Deutschlands (***SPD***) Social Democratic Party of Germany

Sozialdemokratische Partei Österreichs (***SPÖ***) Social Democratic Party of Austria

Sozialistische Arbeiterpartei Deutschlands (***SAP***) Socialist Workers Party of Germany

Sozialistische Arbeiterjugend (***SAJ***) Young Socialist Workers, the youth organisation of the ***SPD***

Stahlhelm Association of front soldiers created after WWI

Sturmabteilung (***SA***) Storm Troopers, a Nazi paramilitary organisation

Universität university
Urteil judgement
Vaterländische Front Fatherland Front, established by Austrian Chancellor Dollfuß in May 1933, advocating Austrian nationalism and independence from Germany
Verbindungsmann intermediary or contact, a term used within communist underground cells
Verbindungsmann im Gebietsmaßstab contacts at the district level, a common term within communist underground cells
Vertrottelt senile
Völkisch nationalist or populist. The term has no direct translation in English but is connected with race and ethnicity
Völkischer Beobachter People's Observer, the official newspaper of the Nazi Party
Volksdeutsche ethnic Germans in other countries
Volksfeinde enemies of the people
Volksfront People's Front, proposed by the ***KPÖ*** as part of a broad anti-fascist front in 1937
Volksgerichtshof People's Court, first created by the Nazi Party in April 1934 to deal with political crimes, e.g. high treason and treason
Volksschule compulsory primary school education
Wahlkreise electoral districts. The Weimar Republic had 35 electoral districts in the period 1924 to 1933
Wahlkreisverbände electoral constituency associations. The Weimar Republic had 16 constituency associations in the period 1924–33
Wahlverteidiger a defence lawyer selected by the defendant
Wehrkraftzersetzung undermining the fighting spirit of the German armed forces
Weltanschauung ideology or view of the world
Zentrumspartei Centre Party representing German Catholics
Zersetzungshochverrat subversive high treason
Zuchthaus penitentiary, a prison term involving hard labour and loss of civil rights

Chapter 1
Introduction

The historian Eric Hobsbawm (1994) famously described the short twentieth century (1914–1991) as the 'Age of Extremes', a period characterised by widespread experimentation with new 'modernising' ideologies and forms of society which ultimately led to two world wars and engendered excesses and inhumanities on a scale, the likes of which had hitherto never been seen. One of the most defining features of this age was the rise of a new variety of authoritarianism that sought nothing less than the total control of all aspects of the society over which it presided. At their peak in 1977, these new types of dictatorships held half of the world's population under their sway (Rosen 2017). With the collapse of the Soviet Union in 1991, it was widely assumed that their time had finally passed and that the spread of liberal democracy to all corners of the world was inevitable. Yet, nearly 30 years later, such pronouncements of the imminent death of authoritarianism seem at best premature. Indeed, to some, it is democracy, rather than authoritarianism, that is now in retreat. As the political scientist, Larry Diamond (2016, p. 151), recently wrote in *Foreign Affairs*: 'between 2001 and 2015, democracy broke down in twenty-seven countries…Meanwhile, many existing authoritarian regimes have become less open, transparent, and responsive to their citizens'. Even in the great bastions of liberal democracy, the USA and the UK, many pundits are now openly talking of the decline of former democratic traditions and practices (Allen 2014, p. 8). In purely quantitative terms, too, it is evident that the democratic cause has a long way to go. According to the Economist Intelligence Unit, in 2016 only 4.5% of the world's population lived in 'full democracies' compared to 32.7% in authoritarian regimes. If 'hybrid regimes' like Myanmar and Mozambique, which blend elements of democracy with healthy doses of authoritarianism, are included, as much as 50.7% of the world still live under regimes which are far from being democratic (The Economist Intelligence Unit 2017). The authoritarian state has not gone away nor is it likely to disappear in the foreseeable future.

Given its persistence and prevalence as a form of government, the authoritarian state, thus, commands attention. And, certainly, researchers, especially during the heights of the Cold War, have exerted considerable intellectual effort in coming to

W. Geerling, G. Magee, *Quantifying Resistance*, Studies in Economic History,
DOI 10.1007/978-981-10-6008-3_1

terms with the phenomenon. One of the more dominant approaches to have emerged for this milieu sees more similarities than differences between the great dictatorships of the twentieth century, a finding which, if true, would imply that a single comprehensive theory of totalitarianism may well be sufficient to understand all such regimes, irrespective of their founding ideology.[1] Recent research, however, has tended to move away from both viewing such dictatorial systems as all-powerful monolithic entities ruthlessly and completely controlled by a ruling elite or party and the belief that all dictatorships were uniform in character and function.

A good illustration of this new outlook is the changing perspective among scholars of comparative legal systems on the role and freedom of action of courts in authoritarian states. Dictatorships, it has traditionally been asserted, ever fearful of losing control of events or of providing their opposition with an avenue to express their discontent, are inherently incapable of affording meaningful judicial discretion or independence.[2] Recent studies, however, have begun to paint a more varied picture, in which some regimes, most notably in twentieth-century Latin America and North Africa, seem both willing and able to provide a degree of empowerment to their judiciary. Whether such empowerment is permitted to develop depends on a host of political considerations, not least among them the relative importance to the state of bureaucratic efficiency, institutional stability, political control and foreign investment.[3]

Similarly, the sudden collapse of a string of communist countries between 1989 and 1991 has challenged, and led to the re-evaluation of, the long assumed robustness of those states. Some of this research has sought to identify and understand better the dynamics of the forces of decline inherent in the mature versions of these systems, whereas others have looked elsewhere, especially in the states' early formative periods, for answers.[4] This renewed burst of activity into the history of dead dictatorships is in part attributable to the opening up of closed archives in Eastern Europe, but it is also tied to a desire to understand better how a flawed system could maintain such seeming stability for so long and then how all that could collapse so suddenly and ignobly. Were the demises of these dictatorships inevitable and rooted, as it were, in the DNA of their system, or were there alternative paths that could have been steered? Much of the recent research in this field is founded on the prem-

[1] See, for example, Arendt (1951) and Gleason 1995).

[2] For examples of the traditional view of authoritarian courts, see Tate (1995, pp. 27–37) and Solomon (2007, pp. 122–45).

[3] For research which suggests some degree of judicial autonomy, see, for example, Pereira (2005), Hilbink (2007) and Ginsburg and Moustafa (2008). On some of the potential determinants of judicial autonomy, see Widner (2001), Finkel (2008), Helmke and Rosenbluth (2009, pp. 345–366) and Smith and Farrales (2010, pp. 163–193).

[4] Much of this research is perforce country specific in its focus. Taking the case of East Germany as an example, for analysis of the underlying dynamics of demise, see Kopstein (1997) and Pfaff (2006). As for research on its formative years, see, for example, Malycha (2000); Pritchard (2000); Naimark (1995); Zank (1987); Benedikt (2003) and Allinson (2000). A useful introduction to the literature is given in Lammers (2002, pp. 333–342).

ise that insights and answers to such questions can only be uncovered by detailed historical analyses that go back to the origins of the system.

One aspect of authoritarianism that to date has attracted comparatively less attention, though, is the study of those individuals and groups who fundamentally challenge from within the dictator's claim of total authority. Unless they prove successful in toppling the regime – and the vast majority of such resisters do not realise their goals – the tendency is for their stories to be ignored or, if not totally forgotten, then perhaps in a later, more enlightened era, to be depicted in highly monumentalised forms. In many ways, this sidelining of resisters is a strange omission from the literature on authoritarianism. For one thing, their bravery cannot be denied. Confronting tyranny at home takes great conviction and courage. As it is also highly likely to imperil the lives of friends, family and colleagues, it is not a course of action entered into lightly. Yet, perhaps somewhat surprisingly given the brutality and comprehensiveness of state terror, such resistance is rarely snubbed out. In fact, the histories of most authoritarian regimes are marked by at least some degree of internal resistance, admittedly, at most times, by just a very few.

Nazi Germany, one of the most tyrannical regimes of all time, certainly conforms to this pattern. Between 1933 and 1945, small numbers of Germans and Austrians actively resisted the Nazi regime from within, challenging its legitimacy, undermining its support in the community and offering alternative futures to their nations. This book seeks to add to, and extend, our knowledge of these resisters. By utilising quantitative analysis to shed new light on various dimensions of the serious resister's experience, this work contributes to a growing corpus of research which in recent years has used data from Nazi Germany to better understand a range of economic, social and historical processes including the direct and external effects of Jewish expulsions, the importance of political connections, the role of social capital in the rise of the Nazis, the persistence of anti-Semitic beliefs and the relationship between different forms of capital and scientific creation (Akbulut-Yuksel and Yuksel 2015, pp. 58–85; Ferguson and Voth 2008, pp. 101–137; Satyanath et al. 2017, pp. 478–26); Voigtländer and Voth 2012, pp. 1339–1392; Voigtländer and Voth 2015, pp. 7931–7936; Waldinger 2016, pp. 811–831).

The remainder of this introductory chapter has three parts. In the following section, we begin by surveying the historiography of resistance in Nazi Germany and then discussing the meaning of serious resistance. This is followed in the second section by a consideration of the use, and potential, of quantification to the study of resistance. Finally, the chapter concludes with a preview of what can be expected from the chapters that lie ahead.

1.1 Understanding Resistance

Since the end of the Second World War, the story of resistance to Nazi tyranny has played a pivotal role in how post-war Germany and Austria have come to understand their past, present and future. It has proved – and continues to prove – neither a simple nor uncontested process. In many ways, it has been a journey whose course has both reflected and been shaped by a need: the need to explain, justify and rationalise the divisions and tensions of both the Cold War and post-Cold War environments.[5]

For the former German Democratic Republic (GDR), the leadership and direction provided by the Communist Party of Germany (*Kommunistische Partei Deutschlands* or *KPD*) to the broader anti-fascist resistance movement of the Nazi years were central to both its foundation myth and its continued *raison d'être*. Not surprisingly, studies of resistance that emanated from the GDR tended to be rather constrained in their scope, portraying resistance as a struggle against imperialism and finance capitalism and focusing almost exclusively on the actions of the *KPD* (Dorpalen 1985; Reich and Finker 1994, pp. 126–142). From the regime's perspective, such a portrayal had a dual purpose: to instruct the East Germans about the horrors of the past and, by stressing the presence of capitalist imperialism in Western society, also educate and warn them about, what their rulers perceived as, the ongoing nature of the class struggle (Kershaw 2015, p. 14). The underground resistance of the *KPD* was the starting point for nearly all historical studies; other groups were ignored or presented in a negative light, as a document from the East German government entitled: 'Improving Research and Teaching in History in the GDR' (*Zur Verbesserung der Forschung und Lehre in der Geschichtswissenschaft der DDR*) testifies:

> In view of the leading role of the working class in the national struggle for liberation, the utmost importance is to be given to research and presentation of the revolutionary traditions of the German workers movement, especially the party of the German working class [Communist Party] and the destruction of the imperialist and social democratic distortion of the history of the German working class struggle. (Reich and Finker 1994, p. 133)

By contrast, in the West, historians, at least initially, concentrated on conservative, bourgeois and military figures, not only because many of the best-known resistance

[5]The immediate post-war period, before the advent of the Cold War, was the exception. As Hans Rothfels (1949, p. 15), Professor of Modern History at the University of Chicago, and one of many prominent academics of Jewish origin forced to emigrate from Nazi Germany, lamented in his 1948 book *Die Deutsche Opposition gegen Hitler*: 'at the time it was an almost universally accepted view in this country [the USA] that there was not and never had been a German opposition to Hitler worth speaking of; that the Germans, differently from any other people and because of either inborn wickedness or an acquired habit of obedience or a specifically obnoxious political philosophy, voluntarily adhered or meekly submitted to a tyrannical regime of gangsters; that they deliberately closed their eyes to horrible crimes committed by Germans'. This value judgement attributed 'collective guilt' to the entire adult German population between 1933 and 1945, implying moral blame for Nazi crimes. Germans were forced to come to terms with certain elements of the past, specifically the legacy of Nazism and the Holocaust, through the Nuremberg War Crimes Trials and de-Nazification proceedings. These events occupied the attention of politicians, historians and the general public alike in the period preceding and shortly after the publication of Rothfels' book. It is not surprising that in this climate the issue of German resistance against National Socialism was largely ignored, as it sat rather uncomfortably alongside the key values perceived as essential in the rehabilitation of the German nation: atonement, punishment and compensation.

figures came from these circles but also because they could establish a link, no matter how strained, between the most evident attempt to remove Adolf Hitler, the 20 July bomb plot, and core values of the Federal Republic of Germany (FRG): freedom and democracy.[6]

In line with the prevailing description in the West of Nazi Germany as a totalitarian state ruled by a monolithic controlling power, the earliest notions of resistance in the FRG singled out a few select groups as heroes and martyrs (Breyvogel 1993, p. 13). So much so that the picture of resistance that had emerged by the early 1950s was fundamentally one dimensional, excluding multiple perspectives and concentrating almost solely on conspiratorial plots epitomised through the actions of Claus Schenk Graf von Stauffenberg. This focus was, in part, due to the limited information available, whether because of missing records, sealed reports or loss of evidence and witnesses. Yet, other more important intentions arguably played a significant role, too. A diversified picture of resistance that included the activities of communists may have induced some people in West Germany to question the activities of the ruling elites in the period leading up to 30 January 1933 and during the Third Reich. Such an examination may have undermined the legitimacy of the conservative elites to govern. Therefore, recognition of conservative-national elites in the military as vanguards of a noble German tradition helped safeguard this position of influence and legitimacy to govern. Bernd Rusinek identifies this subtle repositioning of characters in the picture of resistance:

> In the prioritising of the conspirators of 20 July, one now recognised a privilege of the upper classes. Only the resistance 'from above' had been deemed worthy of recognition; the establishment could always recognise only itself as a historical power, while it said nothing about resistance from workers or marginal groups in society (Rusinek 1994, p. 295).

This political and ideological bias of governments in both of the newly formed German states was also an important factor in defining the scope of academic historical writing in this period. Both the GDR and the FRG attempted to influence the political consciousness of their people and saw promotion of 'resistance' as a legitimisation of their respective states and political systems. Peter Hoffmann (1988, p. 2) notes that when resistance literature first appeared in West Germany, it re-evaluated resistance as a revolt of conscience (Leber et al. 1968). The publication of memoirs from those involved in the 1944 assassination plot was used to counter the collective guilt of Germany by providing the German people and their former enemies with an insight into the 'other Germany'. Resistance fighters were presented as patriots not traitors (Kershaw 2015, p. 217). Chancellor Konrad Adenauer, the first West German Chancellor, felt that it was the duty of every German to protect and save from defamation 'those who with feelings of duty to morality and their fatherland, took any measure to save Germany or at least lessen the proportion of catastrophe, into which the Nationalist Socialist leadership had knowingly plunged Germany' (Ueberschär 1994, p. 10).

---

[6] Mommsen (1966, pp. 75–76), for example, has shown that the ideas of national-conservative resisters were not inspired by the ideals of democracy but rather could be traced back to the pre-Weimar period.

This elevation of specific groups and individuals to mythological status in both post-war German states diverted attention away from the general populace's prior inability or refusal to resist National Socialism, serving to legitimise complicity, passive acceptance or opportunistic behaviour. The moral courage and idealism of the Stauffenberg conspiracy were to be applauded, yet the failure of the assassination attempt on Adolf Hitler and the conspirators' subsequent denunciation as traitors offered a somewhat different message that could be used as a rationalisation for West Germans who wished to forget or excuse their own behaviour: resistance against the National Socialist state was virtually impossible, brought consequences and was ultimately futile unless one was prepared to risk one's life. In this way, the inactivity and guilt of the population could be explained away as a normal reaction to the fear of being perceived to have committed high treason, an excuse which nicely complemented the other common trope that 'we could not resist'.

A different type of rationalisation was employed in the GDR. German fascism had been a mass movement manipulated by finance capital; its defeat in 1945 was a triumph for socialism.[7] A series of biographies heaped praise upon the heroic anti-fascist resistance of the communists. Official propaganda dictated that the population of the GDR was on the victorious side. Freed from the tyranny of fascism, recriminations could be directed outward, specifically towards the Western societies which still harboured fascist elements, allowing the GDR population to avoid coming to terms with its past (Bialas 1993, p. 553).

Starting from such narrow, highly politically charged and polarising positions that afforded overwhelming prominence to specific, 'monumentalised' groups of resisters at the expense of all others, historical understanding of resistance, at first just in the West, and now in a reunified Germany, has thankfully, if gradually, developed and broadened in both approach and focus. This process began in the FRG in the 1960s when a younger generation of historians started to challenge previously accepted historical understandings of German resistance (Rusinek 1994, p. 294). Aided by the discovery of new material, previously unseen records and reports, and writing in a climate dominated more by radical social and political change than moral outrage over events that transpired during the war, historical scholarship based on social and political science began to permeate German universities.

As a result, the nature of scholarship on Nazi Germany changed. The starting point of earlier research had been in accordance with the prevailing description of Nazi Germany as a totalitarian state, but according to Bernd Rusinek (1994, p. 294), 'younger historians directed their attention no longer to the suffering-passive side, but rather to the intensely active side, whereby they demonstrated that possibilities for struggle against an unlawful regime had indeed existed'. This was especially evident among the German working class. Communist resistance, which to this point had been ignored by Western historians, found favour within a less authoritar-

[7] The linking of anti-fascist resistance with anti-capitalist values was not based on hard historical analysis. Indeed, communist resistance against Nazism appears to have had little to do with the roots of fascism in bourgeois society; see Wippermann (1998, pp. 269–70).

ian political environment. Detlev Peukert (1976) was the first prominent Western historian to promote studies of working class and communist resistance with the publication of 'Ruhr Workers against Fascism: Documentation of Working Class Resistance in the Ruhrland 1933–1945' (*Ruhrarbeiter gegen den Faschismus. Dokumentation über den Arbeiterwiderstand im Ruhrgebiet 1933–1945*). This study and a series of 'Resistance and Persecution' (*Widerstand und Verfolgung*) publications by Kurt Klotzbach (1969), Hans-Josef Steinberg (1969) and Kuno Bludau (1973) can be seen as a reproach to the elder generation of historians and the political establishment which held sway in the period before 1968.

Accordingly, the concentration on resistance from above, that is, by the elites and nobility, gave way to an interest in certain types of day-to-day behaviour or social history, the main focus of which concerned the analysis of the actions, attitudes and behaviour of ordinary Germans. Resistance historiography came to be seen as 'social history from below' or *Alltagsgeschichte* (Kershaw 2015, pp. 253–254). Towards the end of the decade, new research emerged centring on regional and local case studies, in particular working class and student resistance and the fate of minority groups such as Jehovah's Witnesses, foreign workers, victims of euthanasia and homosexuals.

One piece of scholarship more than any other, 'The Bavaria Project on Resistance and Persecution in Bavaria 1933–45' (*Das Forschungsprojekt Widerstand und Verfolgung in Bayern 1933–45*), begun in 1973, signalled the beginning of a new era in resistance historiography (Jaeger and Rumschöttel 1977; Broszat et al. 1977–1983).[8] This era, which remains very much in play to this day, has been underpinned by a more encompassing view of resistance that included every form of active or passive behaviour that rejected either totally or some aspect of the National Socialist regime or its ideology and claims to power. It is an approach that has placed partial, passive, ambivalent and sporadic opposition that sought only to limit or immunise individuals or groups against Nazi intrusion – labelled *Resistenz* by Martin Broszat – alongside fundamental acts of resistance (Broszat 1991, p. 25).[9]

In Austrian historiography, too, a similar trend towards a broader and more inclusive understanding of resistance had become evident by the mid-1970s. In 1975, the Documentation Centre of Austrian Resistance (*Dokumentationsarchiv des österreichischen Widerstandes* or *DÖW*) published 'Resistance and Persecution in Vienna 1934–1945' (*Widerstand und Verfolgung in Wien 1934–1945*), the first in a long series of research projects that would eventually extend to cover every province in Austria (*DÖW* 1975; *DÖW* 1979; *DÖW* 1982; *DÖW* 1984; *DÖW* 1987; *DÖW* 1991). The focus of the series was, as the introduction to the first edition explained, not just, as before, on those who had fought for an independent and democratic Austria, but 'the whole range of resistance, opposition, dissatisfaction, that is, every non-

[8] The Bavaria Project consisted of six volumes.

[9] Broszat (1986, pp. 68–91) highlighted the shortcoming of 'resistance' as defined in early post-war writings: 'The definition of exceptional cases [fundamental and active opposition] produced an idealised and undifferentiated picture of German resistance'.

conformist reaction to the dictatorship' (*DÖW* 1975, p. 10–11).[10] In a sense, in adopting such a focus, the *DÖW* was building upon a foundation laid in earlier research by Karl Stadler (1966), who in 'Austria 1938–1945 as Reflected in National Socialist Documents' (*Österreich 1938–1945 im Spiegel der NS-Akten*) had concluded that in view of the absolute demand for obedience made by those in power and the ever-present threat of sanctions, every sporadic act of opposition in the Third Reich must be treated as an act of resistance. Thus, in Austria as in Germany, the placement of all forms of active and passive resistance together on a level footing has become the norm of historical discourse.

Where controversy still exists among mainstream historians of German and Austrian resistance, it is usually confined to theoretical issues associated with the clarification of, and distinction between, the key concepts: resistance, *Resistenz*, non-conformity and so on.[11] Of greater importance is the widespread agreement among historians and scholars of Nazi Germany that greater emphasis needs to be placed on everyday acts of disobedience in order to provide a more balanced, complete picture of resistance to National Socialism.[12]

In 1983, the FRG's President, Richard von Weizsäcker, commissioned historian Peter Steinbach with the task of erecting a permanent exhibition dedicated to all manifestations of German resistance to National Socialism. Steinbach (1994, p. 15) felt that a permanent exhibition should meet certain criteria. Firstly, it should present the entire opposition to National Socialism in its broad and diverse forms. Every attempt should be made to avoid the subjective interpretation of goals as good or anti-totalitarian or to dismiss some as anti-democratic or totalitarian. A pluralistic society with varied conceptions of history should in no way permit such value judgements. Secondly, the memorial to resistance needed to reflect a collective and unifying view of German history, rather than be used as a tool to justify or legitimise states or political systems through the selective representation of particular groups (as had been common pre-1989). All individuals, groups and organisations, irrespective of their background, ideology and goals, including what Steinbach (1994, p. 228) has called the 'unsung heroes' (*Unbesungene Helden*) – those forgotten individuals, who had been relegated to the periphery of a discussion restricted to conspiratorial revolts directed against the prevailing system – were to be given a place in this exhibition.[13]

---

[10] By writing about the resistance and persecution of all Austrians – irrespective of political, ideological, religious, national or other reasons – the work of the *DÖW* afforded belated attention to the plight and resistance of Jews, religious groups, ethnic minorities, youth, military resisters, individual resisters and groups which had hitherto received virtually no recognition in the Austrian historiography.

[11] Mallmann and Paul, for example, seek to replace the term *Resistenz* with reluctantly loyal; see Mallmann and Paul (1993, pp. 99, 105).

[12] Kershaw (2015, p. 239), Koonz (1991, pp. 49–51), Large (1991, p. 4) and Peukert (1991, pp. 36–38) suggested that any study of resistance includes recognition of all acts of non-conformity.

[13] Steinbach (1994, p. 228) writes: 'The deserter is proof that within the strong hierarchy of 'armed power', there were alternatives to obeying orders; on the other hand, the helper of endangered Jews

Through its inclusivity, the emerging view sought to build consensus or at the very least avoidance of a return to the state of conflict-driven discourse. As the preface of the 1997 *Encyclopedia of German Resistance to the Nazi Movement* pointed out, its contributors shared an outlook that was 'based on a broad notion of resistance that must include every form of disobedience and opposition, from quiet obstructionism to the attempted murder of a tyrant. The editors, and authors' goal was to do justice to all forces that opposed National Socialist ideology and rule' (Benz and Pehle 1997, p. xi).[14]

Above all, it is a consensus based on the absence of judgement. No verdicts are made, and the relative merits and importance of different strains of resistance are rarely, if ever, explicitly discussed. In a sense, it is as if all forms of opposition are levelled and then lumped together into an undifferentiated, albeit fragmented, collective mass.[15] As an approach, it represents the antithesis of the competing, monodimensional stances that prevailed at the height of the Cold War and is, in fact, precisely the reason why in many historical accounts communist resisters can now sit at least moderately comfortably side by side with national-conservative. It is an approach redolent of the more inclusive, tolerant and democratic post-*Wende* (following the fall of the Berlin Wall) times. For, as Benz (2014, p. 119) has recently noted, 'in an historical culture, which is orientated towards justice and follows objective criteria, there must be room for all streams of resistance against National Socialism: for those with bourgeois-conservative roots like the 20 July group, as well as the International Socialist Militant League and those others drawn from the tradition of the working class resistance groups including the Communists'.

Not all have embraced this objectivity and pluralism, of course. Some reject the moral relativism of this approach, in particular its reinstatement of communists back into resistance historiography (Nolte 1987, p. 440). Yet, such views are in the minority. For many, retaining consensus is important. As Wolfgang Wippermann

---

shows that even in everyday life, the web of the 'totalitarian state' does not provide an excuse for refusing to help or showing compassion'. Steinbach's focus on everyday acts of disobedience reflected the trend of the literature to go beyond 'forgotten' political resisters to include other groups marginalised in studies of resistance. In their article *Das andere Deutschland*, Grebing and Wickert (1994, p. 222) contended that 'the goal of further research into resistance and persecution of women and men should be aimed at tracking down the ambivalences in their lives, which for women arose from the contradiction between family tasks and occupational work, the possibility of resistance activities and survival in the period of National Socialism'.

[14]A similar expression of the broader notion of resistance is reflected in the current *Lexikon des Deutschen Widerstandes*, edited by Peter Steinbach and Johannes Tuchel in 1998. 'In this connection, resistance denotes all active and passive conducts directed against the National Socialist regime or a considerable aspect of the National Socialist ideology which involved high personal risks'; see Steinbach and Tuchel (1998, pp. 241–42).

[15]In the words of Gerd Ueberschär (1984, p. 15): 'The 20 July 1944 and other days from the history of resistance can be bridges on the way toward a new collective picture of the National Socialist period and resistance in the unified Germany. This will, however, only succeed when the oppositional behaviour of 'the simple people', the small forms of refusal in everyday life are given greater emphasis in the evaluation of resistance than previously, when one tended to place too much emphasis on certain prominent personalities and elites.

(1998, p. 277) has opined, attempts at 'such exclusion and devaluation of important parts of the resistance, based exclusively on party-political and in no way scholarly motives, threaten the earlier consensus to honour the resistance of all groups'.

There can be little doubt in the appropriateness of reporting and respecting the bravery of all who stood up to tyranny: each of their narratives deserves to, and should, be told. Yet, by the same token, one must concede the possibility that in allowing a multitude of voices to be heard on an equal basis, one might also conceal distinctive features of certain voices behind the overall wall of sound. Does, for example, placing all acts and forms of resistance on the same plain of historical importance engender a sense of consensus at the expense of obfuscating historical understanding? While the experience of every resister is important, not just to themselves, their families and communities but also for its part in the building of a richer picture of the period, do we not push this view too far when we assume, implicitly or explicitly, that everybody's experience has a concomitant impact on the broader unfolding of the historical narrative? Certainly, avoidance of the issue of impact will not provide answers.

There are two problems with assessing the relative impacts of resistance activities. First, there is a possibility that the exercise will lose objectivity and be manipulated for political purposes, in the process polemicising the debate and demeaning certain groups and communities. Given the historiography of resistance, one has reason to express such a concern. Second, there is the question of measurement: How can an impact be fairly and consistently assessed?

No doubt, these concerns pose difficult challenges, and we will take them up again in more detail in the concluding chapter of this volume. Yet, as Mary Fulbrook (1995, p. 288) has reminded us: 'it is nevertheless the task of the historian — however personally sympathetic to one cause or another, however hostile to the views or activities of others in the drama — to take a step back and seek to present a sober analysis of all the relevant factors as they are captured in the particular net of categories and concepts'. She was referring to the writing of the history of the GDR, but the point applies equally to the historiography of German resistance. In his discussion of the definition of such resistance, Ian Kershaw (2015, p. 214) made a similar appeal for researchers to be guided by objectivity noting that pursuing such an approach with care need not necessarily lead to feelings of disrespect being felt in some quarters: 'But respect [for those who stood out, in whatever ways, against repression and inhumanity] must not hinder the task of rational, critical assessment of the opposition to Nazism – as long as such an assessment keeps in mind the 'art of the possible', and does not attempt to pass judgement on the basis of ahistorical moral ideas'.

This book seeks to adopt such a rational, critical approach to the question of the impact of resistance. As the forgoing discussion of the historiography of resistance has emphasised, the meaning of 'resistance' differs radically between authors and places. The term is now commonly used to embrace a wide range of behaviours, spanning from the telling of an anti-regime joke or the failure to give the Hitler salute right through to attempts to bring down the regime (Möller 2013). While recognising that there were many ways to resist or oppose the regime and that individual actions were perforce shaped by the context they found themselves in and what they could realistically do, in this work we employ the term 'resistance' in its much narrower, perhaps more traditional, sense.

In making sense of the array of terms offered by historians to conceptualise German opposition to Hitlerism, Detlev Peukert (1991, pp. 35–48) has devised a useful and realistic framework whereby resistance and non-conformist behaviour are two endpoints on a sliding scale of dissident behaviour. By using a scale of complexity and risk, he distinguishes among types of conflict, beginning with occasional private non-conformism, confined to the private sphere, proceeding to wider acts of refusal, involving some effect on public opinion, and then to protest, where the act is highly visible and contains some intentional effect on public opinion. In this schema, the term 'resistance' is confined to a form of behaviour where the intention is to make a public impact and pose a basic challenge to the regime.[16]

We accept the basic premise of Peukert's approach but believe that for an act to be labelled 'resistance', in addition to having a public impact and posing a basic challenge to the regime, it must involve a principled rejection of the regime based on political-ideological or moral convictions. We, therefore, place greater emphasis on motivation than some others in the literature. Moreover, in this work, we focus our attention on what we call 'serious resistance', namely, resistance as defined above, which the Nazi state itself viewed as constituting a potential challenge to its survival. As explained more fully in Chap. 2, we employ court records to determine which resistance activities met this definition of seriousness. In line with Ian Kershaw (2015, p. 239) and Claudia Koonz (1991, p. 49), we favour a category of opposition to cover all types of non-conformist behaviour. The term *Resistenz*, which attributes significance to any act that undermined National Socialist rule (regardless of the motive), fails to meet our criteria for serious resistance. As a concept, it appears better suited to the wide spectrum of everyday non-conformist behaviour in Nazi Germany that, despite its commonality, did not in itself seek to bring down the regime.

1.2 Quantifying Resistance

As a glance forward to the pages that lie ahead reveals, one of the distinctive features of the research reported in this book is its conscious attempt to capture quantitatively various dimensions of both the phenomenon of resistance itself and those who actively took part in it. We believe that quantification and statistical

[16] It must be said, however, that the advantage that Peukert's framework provides in making a clear distinction between forms of behaviour is partially offset by the lack of conceptual clarity. The qualitative difference between acts of non-conformism, refusal and protest is not always easily definable. Nor is it clear how the parameters of complexity and risk delineate a clear boundary both within these forms of behaviour and between non-conformism and resistance. Does protest invariably lead to resistance?

analysis – when read closely in conjunction with the literature, employed with due care for historical context and interpreted with full understanding of the limitations of the sources utilised – offer tools to the historian that can help not only sort out questions of scale and magnitude and separate fact from myth but also identify hitherto unnoticed vistas of research. Moreover, we contend that forming such a comprehensive overview of historical fact is in any case a necessary first step in any rational and objective analysis of resistance. As Brigitte Bailer (2013, p. 90) has recently reminded us in her work on the Austrian victims of National Socialism, 'to quote concrete numbers means to make irrevocable and recognise, to advance the history of persecution in the world of measureable facts'.

It is fair to say that quantification has to date not featured particularly prominently in the historiography of German resistance to National Socialism. The overwhelming majority of research on resistance undertaken in Germany has been qualitative in approach, based on the skilful and often highly insightful workings of an array of primary sources. Memoirs, diaries, oral accounts as well as findings gleaned from the close reading of the archives of public and private organisations and institutions have, thus, formed the basis of much of our current understanding. Of course, in making this observation, it is not our intention to imply that quantification – viz. the use of numbers – has previously been completely ignored. On the contrary, we acknowledge that many researchers have used statistics before to amplify their points. Detlev Peukert (1980) and Allan Merson (1985) among others, for example, have frequently deployed numbers to illustrate succinctly the extent of communist and working-class resistance in Nazi Germany. The Topography of Terror Foundation (*Stiftung Topographie des Terrors*) has compiled a list of known inmates of the 'House Prison' at Gestapo Headquarters in Berlin between 1933 and 1945 and used it to draw interesting conclusions about the composition of those prisoners (Topography of Terror Foundation 2014, pp. 214–237). Similarly, Christl Wickert (1994, pp. 200–225) has employed a random sample of 2000 Gestapo files and 400 court files from the Special Courts (*Sondergericht*), Higher Regional Courts (*Oberlandesgericht*) and People's Courts (*Volksgerichtshof*) in Duesseldorf and Essen to examine female resistance, arguing inter alia that women were more likely to become involved in everyday acts of dissent, disobedience or ideological dissent than political resistance per se. These examples could, of course, be replicated many times over. Our point is that, where it has occurred, such quantification has rarely gone beyond the reporting, often en passant, of summary statistics, sometimes to support a contention and other times merely as a piece of interesting information. The raw data underlying those summary statistics are typically not systematically categorised. Nor are they used as a basis for deeper statistical analysis. Quantification in these senses has played at most a marginal role in the historiography of resistance.

Perhaps the biggest exception to this characterisation of the resistance literature – and one of the best illustrations of the potential for quantitative research in the field – is the work published in 2013 by the *DÖW* under the title of 'The Fate of Martyrs: Resistance and Persecution in National Socialism' (*Opferschicksale: Widerstand und Verfolgung im Nationalsozialismus*) (*DÖW* 2013). In that compendium, a number of chapters provide detailed statistical accounts of different aspects of the Austrian martyrs of National Socialism. Two chapters, for example, one by Brigitte Bailer and Gerhard Ungar (2013, pp. 111–124) on deaths as a result of political persecution and the other by Brigitte Bailer, Elisabeth Boeckl-Klamper, Wolfgang Neugebauer and Thomas Mang (Bailer et al. 2013, pp. 163–190) on the Gestapo as an instrument of Nazi terror, use a sample of 7974 victims of political persecution collected from the daily reports and index cards of the Gestapo Headquarters in Vienna. These data are used to shed light on the age, gender and regional structure of Austrian victims over the whole period of National Socialist rule. Annual series for the number of reports filed and index cards created are presented and a breakdown of arrests by categories also made. The 11 categories utilised by the authors represent a curious amalgam of offence types, personal characteristics and political affiliations: working-class movements (communist, social democrats and trade unionists), offences against the war economy (e.g. black marketers), prisoners of war and forced labourers (typically arrested for forbidden contact with local women or arriving late at work), individual resisters (including those accused of malicious gossip), Jewish, Catholic-Conservative organisations, Catholic clergy, Jehovah's Witnesses, homosexuals, members of the Nazi Party and its ancillary organisations and saboteurs. Another chapter by Wolfgang Neugebauer (2013, pp. 233–272) analyses the cases of 6300 Austrian resistance fighters who appeared before the People's Court and the Higher Regional Courts of Vienna and Graz. It reports on the overall regional, age and gender structure of these individuals. Where known, the political affiliation of resisters is broken down into eight groups (Social Democratic Party of Austria and its suborganisations, the Communist Party of Austria and its suborganisations, Catholic-Conservative organisations, Legitimist groups, the Christian Social Party of Austria, Nazi Party members, Jews and Jehovah Witnesses) and the share of each in the total number of cases given for the periods either side of 1934 (Neugebauer 2013, p. 239).

German legal historians have also made a vital contribution to the quantitative history of resistance. Historians of the People's Court, certainly since Walter Wagner's (1974) seminal work of 1974, have generated a wealth of quantitative information on this most politicised of courts. Most of the data collated have come from samples extracted directly from the People's Court's own records or official legal publications of the time. Klaus Marxen (1994) has, thus, gathered information about the trials of 5832 individuals who appeared before the People's Court and Holger Schlüter (1995) for 1371. Working together, Marxen and Schlüter (2004) have also documented and analysed a further 50 previously unpublished People's Court verdicts. Another historian, Edmund Lauf (1994), combed through the press work of the People's Court and some 231 articles about the Court featured in the Nazi Party organ, the 'People's Observer' (*Völkischer Beobachter*), to give a broad statistical overview of nearly 15,000 defendants. Others have utilised the same sources to analyse cases and defendants from particular locations. Wolfgang Form and Rolf Engelke's (Form 2007; Form and Engelke 1996, pp. 26–43) research, for example, has studied the Supreme Court and People's Court trials of individuals from Hessen, Darmstadt and Frankfurt am Main.

These works have mainly explored questions of legal process and the functioning of the courts in question. Not surprisingly, they tend to provide statistics on matter such as the numbers of cases heard, defendants seen, charges laid and the frequency with which different sentence types were handed down. They also frequently include some broad categorisation of defendants by political, ethnic or religious background, as well as some notion of periodisation. By contrast, next to nothing on the profiles, activities or characteristics of the defendants themselves are given. Their overwhelming focus is, thus, on the workings of the Nazi court system, not resistance per se. As a result, the degree of cross-fertilisation between resistance and legal historians has been rather limited. Typically, historians of resistance cite the works of legal historians only when they wish to reference the number of cases heard or death sentences handed down in a particular period.

In addition to the legal historian's different focus, students of German resistance have been discouraged from drawing upon these quantitative sources by the fact that many of the works in question are perceived to have been constructed on statistically convenient, rather than historically robust, grounds. Jürgen Zarusky (2011, p. 1012), thus, has criticised Holger Schlüter for his overreliance on statistical materials at the expense of historical context and the 'self-generated' nature of the results produced as a consequence. An illustration of the type of concern raised by Zarusky is the categorisations employed by legal historians. In some works, the classifications and aggregations presented appear to have been chosen primarily by their ready availability in the source material. Typically, no attempt is made to disaggregate and then recombine the data (if possible) into more meaningful classifications or to extend what is found in the court records with data extracted from other sources. One consequence is that a lot of the data available are left underutilised. Thus, to take just one example, the organisational affiliation of as much as 41.7% of Edmund Lauf's (1994, p. 240) sample of defendants could either not be characterised at all (23.4% of the total) or only as part of a broad catch-all category called 'other' (18.3%).

This problem is compounded by the, at times, rather idiosyncratic nature of the categories adopted, classificatory schemes that, frankly, do not seem to be informed, let alone driven, by broader historical debate or even the works of others in the same field who have gone before. Such schemes simply render the comparability, interpretation and, indeed, meaningfulness of the findings of different researchers all the harder. Two areas of interest – nationality and resistance groups – prove particularly problematic in this regard. Klaus Marxen's (1994, pp. 30–35) analysis of nationality is complicated by definitions of *Volksdeutsche* and *Reichsdeutsche*, which are at odds with usual conventions within the literature. He classifies *Reichsdeutsche* as Germans who lived within the pre-1938 boundaries and *Volksdeutsche* as ethnic Germans with foreign citizenship, including, unusually, Austrians and Sudetens. Edmund Lauf (1994, p. 232) adopts an even simpler categorisation, distinguishing only between Germans and non-German, the latter category including Austrians. As Jürgen Zarusky (2011, p. 1012) has noted, categorising Austrians and Sudetens as either *Volksdeutsche* or non-German is problematic from a conceptual perspective as each of these groups was in a legal sense treated by the German Reich in the same manner as ethnic Germans born within its pre-1938 boundaries. Given their historical, linguistic and cultural similarities, it makes sense to include Austrians and Sudetens between 1938 and 1945 in the category *Reichsdeutsche.*

The process of classifying resistance groups has invoked an even greater diversity of approaches. Klaus Marxen's (1994, p. 40) five-part classification of German resistance groups – (i) left wing, (ii) undermining the fighting spirit of the German armed forces (*Wehrkraftzersetzung*), (iii) national conservative, (iv) espionage and (v) others – appears rather perfunctory. The considerable breadth of these groupings and their mixing together of internal political and international motivations with a diverse range of individual, often off-the-cuff, reactions to the war are simply inadequate to capture with any degree of granularity the differing patterns, temporally and in deed, taken by specific resistance movements. Lumping together, the resistance of the *KPD*, Social Democratic Party of Germany (*Sozialdemokratische Partei Deutschlands* or *SPD*), Socialist Workers Party of Germany (*Sozialistische Arbeiterpartei Deutschlands* or *SAP*), trade unionist and a plethora of small socialist and anarchist organisations in fact conceals more than it reveals. Similarly, not distinguishing between conservative, *völkisch*, military, catholic and confessing church resistance does little to advance our understanding of resistance from the centre and right of the political spectrum.

Edmund Lauf's (1994, pp. 237, 240) classification of resistance groups is likewise vague and changeable. In the key table, he has six resistance groups; in another table, there are eight. Admittedly, he makes some attempt to decompose the various left-wing groups by ideology and nationality (*KPD*, foreign communist parties and German socialists), but around a quarter of resistance groups remain classified as 'other' and nearly one in five as 'individuals acting alone'. When more than 40% of a sample cannot be easily allocated within a classificatory scheme, it suggests that that scheme is far from exhaustive.

Holger Schlüter (1995, p. 109) arguably provides the most comprehensive approach. He offers the following five-part classification of resistance groups: (i) espionage (treason), (ii) opposition (both left- and right-wing political opposition that has continued on from the Weimar period), (iii) home front (defeatism, giving aid and support to the enemy [*Feindbegünstigung*], listening to foreign broadcasts), (iv) annexation (Austrian, Polish and Czech independence movements) and (v) other (inter alia: economic sabotage as well as French and Belgian victims of Hitler's 'Night and Fog' [*Nacht und Nebel*] directive). Despite its broader reach, from the perspective of understanding German resistance, this classification still poses many problems. First of all, the classifications are not mutually exclusive. A German communist, for example, could easily belong to category (i), (ii) or (iii) and even category (iv) or (v) if his/her activities took place outside Germany (as was often the case). Moreover, there is a high degree of ambiguity and overlap both between and within the categories, a problem that stems from poorly defined, contradictory criteria. Some aspects of the criteria are based on the location of the resisters, others on the type of activity and yet others on ideology. Other than for a single table where the share of political opposition over the entire period is broken into five groups (*KPD*, *KPD* affiliated, *SPD*, other left-wing and others) – a classification which still lumps nationalist, conservative, religious-based and military opposition together under 'other' – no attempt to delineate between the many different political organisations that resisted National Socialism is made (Schlüter, 1995).

In this book, we follow the classificatory approach adopted by one of the most important data-collating initiatives in resistance studies of recent years, the 'Resistance as High Treason' (*Widerstand als Hochverrat*) project. This framework, which is both exhaustive and internally consistent, is rooted in an understanding of serious political resistance as defined by the regime itself and demonstrated by its willingness to lay charges of treason and high treason charges in response (Zarusky and Mehringer 1998, p. 13). This classification's groupings incorporate well-known parties of resistance, for instance, the *KPD*, Communist Party of Austria (*Kommunistische Partei Österreichs* or *KPÖ*) and *SPD*; groups based around major resistance events (such as the 20 July bomb plot); group designations introduced by the resistance historiography itself, for example, the Hübener group; and, in a few cases, significant organisations, of which resisters were active members, such as the National Committee for a Free Germany (*Nationalkomitee freies Deutschland* or *NKFD*). To avoid an unnecessary proliferation of categories, broader collective groups are occasionally adopted to capture a wide variety of often quite small resistance groups which exhibit common defining characteristics (e.g. the *Bündische Jugend*). Resistance activities outside fixed groups are classified as 'not organised' (Zarusky and Mehringer 1998, p. 56). This classification scheme, which has the advantage of not conflating the political basis of the resistance with features of its form, goal and location, is outlined and discussed in more details in Chap. 2.

1.3 What Lies Ahead

Building on the pioneering work of those who have gone before, this book seeks to extend our knowledge of resistance by providing the first systematic overview *from a quantitative perspective* of the most serious political crimes in Nazi Germany.[17] The formal part of that analysis begins in the next chapter (Chap. 2) with a comprehensive description of the data that lie at the heart of our research. There, details of how the database was constructed, the judgement calls made by the authors in its construction, and how key variables should – and should not be – interpreted are provided. The following three chapters then present a thorough quantitative investigation of the when and where (Chap. 3), who (Chap. 4) and why (Chap. 5) of serious resistance. The role and patterns of inter alia 'lone wolf' unaligned resisters, different political organisations, religious groups and ethnicities, regions as well as different socio-economic backgrounds, age and gender are examined. In the

[17] In order to promote the readability and succinctness of this book, the results of the more detailed econometric and statistical analyses undertaken are not reported in the text. Readers interested in seeing these details and the full estimations underpinning claims made in the text, however, are directed via footnotes to references where that information is readily available.

process, a rare perspective on the resisters themselves is put forward, casting new light on the often overlooked human face of those who, through their actions, actively defied Nazi tyranny. In Chap. 6, our attention shifts to how resisters behaved in custody as well as how they were processed and dealt with by the Nazi legal system. These various threads are then brought together in the concluding chapter (Chap. 7), where some of the key implications of our findings for the understanding of serious resistance activities in Nazi Germany are discussed. One new thing that emerges from this analysis is a framework that for the first time allows the objective evaluation of the impact of serious resistance to be made, an approach which with simple modifications could be easily extended to the study of dissent, opposition and non-conformity.

In concluding an introductory chapter such as this, it is often prudent to finish by spelling out explicitly both what a monograph's purpose is and, crucially, *what it is not*. To that end, our intent in writing this book, as mentioned above, is to lay a solid and consistent quantitative foundation for the study of serious resistance by Germans and Austrians to National Socialism and, thereby, provide – it is hoped – a useful, accurate and comprehensive overview of such resistance for both the researcher and student. With regard to what it is not, this book does not purport to be an encyclopaedia of resistance; nor do we see its primary purpose as being to supply detailed accounts of specific resistance events or resisters. There is, of course, much information contained within this volume on each of these things, but readers seeking such a level of detail would in many cases be better served by consulting works with a specific focus on the events or resisters of interest. That said, we are firmly convinced that the higher-level, social scientific, yet historical grounded, approach adopted by the book does add considerable value to the existing historiography, especially in relation to the *overall* phenomenon of serious resistance, insights that, frankly, could not be so easily obtained from more narrowly defined research. We would further contend that the new avenues of research the book opens up make it well worth the read as well.

References

Primary Sources

(A) Archive Files

(i) Widerstand als Hochverrat

Zarusky, Jürgen and Hartmut Mehringer eds. 1998. *Widerstand als Hochverrat 1933–45. Die Verfahren gegen deutsche Reichsangehörige vor dem Reichsgericht, dem Volksgericht und dem Reichskriegsgericht*. Munich: K G Saur. Microfiche edition.

(ii) Dokumentationsarchiv des österreichischen Widerstandes

Dokumentationsarchiv des österreichischen Widerstandes ed. 1975. *Widerstand und Verfolgung in Wien 1934–1945: eine Dokumentation*, vols. 1–3. Vienna: Österreichischer Bundesverlag.

Dokumentationsarchiv des österreichischen Widerstandes ed. 1979. *Widerstand und Verfolgung in Burgenland 1934–1945: eine Dokumentation*. Vienna: Österreichischer Bundesverlag.

Dokumentationsarchiv des österreichischen Widerstandes ed. 1982. *Widerstand und Verfolgung in Oberösterreich 1934–1945: eine Dokumentation*, vols. 1–2. Vienna: Österreichischer Bundesverlag.

Dokumentationsarchiv des österreichischen Widerstandes ed. 1984. *Widerstand und Verfolgung in Tirol 1934–1945: eine Dokumentation*, vols. 1–2. Vienna: Österreichischer Bundesverlag.

Dokumentationsarchiv des österreichischen Widerstandes ed. 1987. *Widerstand und Verfolgung in Niederösterreich 1934–1945: eine Dokumentation*, vols. 1–3. Vienna: Österreichischer Bundesverlag.

Dokumentationsarchiv des österreichischen Widerstandes ed. 1991. *Widerstand und Verfolgung in Salzburg 1934–1945: eine Dokumentation*, vols. 1–2. Vienna: Österreichischer Bundesverlag.

(B) Other

Topography of Terror Foundation ed. 2014. *The "House Prison" at Gestapo Headquarters in Berlin: Terror and Resistance, 1933–1945*, 3rd edition. Berlin: Topography and Terror Foundation.

Secondary Sources

Akbulut-Yuksel, Mevlude and Mutlu Yuksel. 2015. The Long-Term Direct and External Effects of Jewish Expulsions in Nazi Germany. *American Economic Journal: Economic Policy* 7 (3): 58–85.

Allen, James. 2014. *Democracy in Decline: Steps in the Wrong Direction*. Montreal: McGill-Queens University Press.

Allinson, Mark. 2000. *Politics and Popular Opinion in East Germany 1945–68*. Manchester: Manchester University Press.

Arendt, Hannah. 1951. *The Origins of Totalitarianism*. New York: Schocken Books.

Bailer, Brigitte. 2013. Zur Frage nach Quantifizierungen der NS-Opfer. In *Opferschicksale: Widerstand und Verfolgung im Nationalsozialismus*, ed. *Dokumentationsarchiv des österreichischen Widerstandes*, 89–100. Freistadt: Plöchl Druck GmbH.

Bailer, Brigitte, Elisabeth Boeckl-Klamper, Wolfgang Neugebauer and Thomas Mang. 2013. Die Gestapo als zentrales Instrument des NS-Terrors in Österreich. In *Opferschicksale: Widerstand und Verfolgung im Nationalsozialismus*, ed. *Dokumentationsarchiv des österreichischen Widerstandes*, 163–190. Freistadt: Plöchl Druck GmbH.

Bailer, Brigitte and Gerhard Ungar. 2013. Die Zahl der Todesopfer politischer Verfolgung – Ergebnisse des Projekts. In *Opferschicksale: Widerstand und Verfolgung im Nationalsozialismus*, ed. *Dokumentationsarchiv des österreichischen Widerstandes*, 111–124. Freistadt: Plöchl Druck GmbH.

Benedikt, Gavin. 2003. *Gründe für das Scheitern der DDR-Wirtschaft*. Berlin: GRIN Verlag.

Benz, Wolfgang. 2014. *Der deutsche Widerstand gegen Hitler*. Munich: C.H. Beck.

Benz, Wolfgang and Walter Pehle eds. 1997. *Encyclopaedia of German Resistance to the Nazi Movement*. New York: Continuum Publishing Company.

Bialas, Wolfgang. 1993. Antifaschismus in der GDR. Historisch-kritische Aufräumarbeiten. *Das Argument. Zeitschrift für Philosophie und Sozialwissenschaften* 35: 551–570.

Bludau, Kuno. 1973. *Gestapo-geheim! Widerstand und Verfolgung in Duisburg 1933–1945*. Bonn-Bad Godesberg: Verlag Neue Gesellschaft.

Breyvogel, Wilfried. 1993. Jugendwiderstand im Nationalsozialismus. Ein Überblick. In *Widerstand der jungen Generation im NS-Staat: Dokumentation einer Tagung vom 17–19. Juli 1992 im Adam-von-Trott Haus*, 13–32. Berlin: Evangelisches Bildungswerk, Haus der Kirche.

Broszat, Martin. 1986. *Nach Hitler*. Munich: Oldenbourg.

Broszat, Martin. 1991. A Social and Historical Typology of the German Opposition to Hitler. In *Contending with Hitler: Varieties of German Resistance in the Third Reich*, ed. David C. Large, 25–34. Cambridge: Cambridge University Press.

Broszat, Martin, Elke Fröhlich, and Falk Wiesemann eds. 1977–1983. *Bayern in der NS- Zeit*, 6 vols. Munich-Vienna: Oldenbourg.

Diamond, Larry. 2016. Democracy in Decline. How Washington can Reverse the Tide. *Foreign Affairs* 95 (4): 151–159.

Dorpalen, Andreas. 1985. *German History in Marxist Perspective. The East German Approach*. Detroit: Wayne State University Press.

Ferguson, Thomas and Hans-Joachim Voth. 2008. Betting on Hitler: The Value of Political Connections in Nazi Germany. *Quarterly Journal of Economics* 123 (1): 101–137.

Finkel, Jodi. 2008. *Judicial Reform as Political Insurance: Argentina, Peru, and Mexico in the 1990s*. Notre Dame: University of Notre Dame Press.

Form, Wolfgang. 2007. Der politische Widerstand im Spiegel der NS-Justiz in Hessen. In *Politischer Widerstand gegen die NS Diktatur in Hessan: Ausgewählte Aspekte Nr. 46*, ed. Renate Knigge-Tesche. Wiesbaden: Hessische Landeszentrale für politische Bildung.

Form, Wolfgang and Rolf Engelke. 1996. 'Hochverrat' – 'Heimtücke' – 'Wehrkraftzersetzung'. Zur politischen Strafjustiz in Hessen. In *Verfolgung und Widerstand in Hessen 1933–1945*, eds. Renate Knigge-Tesche and Axel Ulrich, 26–43. Franfurt am Main: Eichborn.

Fulbrook, Mary. 1995. *Anatomy of a Dictatorship: Inside the GDR, 1949–1989*. Oxford: Oxford University Press.

Ginsburg, Tom and Tamir Moustafa eds. 2008. *Rule by Law: The Politics of Courts in Authoritarian Regimes*. New York: Cambridge University Press.

Gleason, Abbot. 1995. *Totalitarianism: The Inner History of the Cold War*. New York: Oxford University Press.

Helmke, Gretchen and Frances Rosenbluth. 2009. Regimes and the Rule of Law: Judicial Independence in Comparative Perspective. *Annual Review of Political Science* XII: 345–366.

Hilbink, Lisa. 2007. *Judges beyond Politics in Democracy and Dictatorship: Lessons from Chile*. New York: Cambridge University Press.

Hobsbawm, Eric. 1994. *The Age of Extremes: The Short Twentieth Century, 1914–1991*. London: Abacus.

Hoffmann, Peter. 1988. *German Resistance to Hitler*. Cambridge: Harvard University Press.

Jaeger, Harald and Hermann Rumschöttel. 1977. Das Forschungsprojekt Widerstand und Verfolgung in Bayern 1933–45. *Archivalische Zeitschrift* 73: 209–220.

Kershaw, Ian. 2015. *The Nazi Dictatorship: Problems and Perspectives of Interpretation*. London: Bloomsbury Revelations.

Klotzbach, Kurt. 1969. *Gegen den Nationalsozialismus. Widerstand und Verfolgung in Dortmund 1930–1945*. Hannover: Verlag für Literatur und Zeitgeschehen.

Koonz, Claudia. 1991. Choice and Courage. In *Contending with Hitler: Varieties of German Resistance in the Third Reich*, ed. David C. Large, 49–65. Cambridge: Cambridge University Press.

Kopstein, Jeffrey. 1997. *The Politics of Economic Decline in East Germany, 1945–1989*. Chapel Hill: University of North Carolina Press.

Lammers, Karl Christian. 2002. The Making of the GDR: New Research on its Formative Years and Problems. *Contemporary European History* 11: 333–342.

Large, David C. ed. 1991. *Contending with Hitler: Varieties of German Resistance in the Third Reich*. Cambridge: Cambridge University Press.

Lauf, Edmund. 1994. *Der Volksgerichtshof und sein Beobachter: Bedingungen und Funktionen der Gerichtberichterstattung im Nationalsozialismus*. Opladen: Springer Fachmedien Wiesbaden GmbH.

Leber, Annedore, Willy Brandt, and Karl D. Bracher eds. 1968. *The Conscience in Revolt: Portraits of the German Resistance 1933–1945*. Mainz: Hase and Koehler.

Mallmann, Klaus-Michael and Gerhard Paul. 1993. Resistenz oder loyale Widerwilligkeit? Anmerkungen zu einem umstrittenen Begriff. *Zeitschrift für Geschichtswissenschaft* 41: 99–116.

Malycha, Andreas. 2000. *Die SED. Geschichte ihrer Stalinisierung 1946–1953*. Paderborn: Ferdinand Schöningh.

Marxen, Klaus. 1994. *Das Volk und sein Gerichtshof: eine Studie zum Nationalsozialistische Volksgerichtshof*. Frankfurt am Main: Klostermann.

Marxen, Klaus and Holger Schlüter. 2004. *Terror und "Normalität": Urteile des nationalsozialistischen Volksgerichtshofs 1934–1945. Eine Dokumentation*. Duesseldorf: Justizministerium des Landes Nordrhein-Westfalen.

Merson, Allan. 1985. *Communist Resistance in Nazi Germany*. London: Lawrence Wishart.

Möller, Lenelotte. 2013. *Widerstand gegen den Nationalsozialismus*. Wiesbaden: Marixverlag.

Mommsen, Hans. 1966. Gesellschaftsbild und Verfassungspläne des deutschen Widerstandes. In *Der deutsche Widerstand gegen Hitler, Vier historisch-kritische Studien*, eds. Walter Schmitthenner and Hans Buchheim, 75–76. Cologne: Kiepenheuer & Witsch.

Naimark, Norman M. 1995. *The Russians in Germany: A History of the Soviet Zone of Occupation, 1945–1949*. Cambridge: Harvard University Press.

Neugebauer, Wolfgang. 2013. Der österreichische Widerstand 1938–1945. In *Opferschicksale: Widerstand und Verfolgung im Nationalsozialismus*, ed. *Dokumentationsarchiv des österreichischen Widerstandes*, 233–272. Freistadt: Plöchl Druck GmbH.

Nolte, Ernst. 1987. *Der europäische Bürgerkrieg 1917–1945: Nationalsozialismus und Bolshewismus*. Frankfurt am Main: Propyläen Verlag.

Pereira, Anthony. 2005. *Political (In)justice: Political Authoritarianism and the Rule of Law in Brazil, Chile, and Argentina*. Pittsburgh: University of Pittsburgh Press.

Peukert, Detlev. 1976. *Ruhrarbeiter gegen den Faschismus. Dokumentation über den Arbeiterwiderstand im Ruhrgebiet 1933–1945*. Frankfurt am Main: Röderberg-Verlag.

Peukert, Detlev, 1980. *Die KPD im Widerstand, Verfolgung und Untergrundarbeit an Rhein und Ruhr 1933–1945*. Wuppertal: Hammer.

Peukert, Detlev. 1991. Working-Class Resistance: Problems and Options. In *Contending with Hitler: Varieties of German Resistance in the Third Reich*, ed. David C. Large, 35–48. Cambridge: Cambridge University Press.

Pfaff, Stephen. 2006. *Exit-Voice Dynamics and the Collapse of East Germany: The Crisis of Leninism and the Revolution of 1989*. Durham: Duke University Press.

Pritchard, Gareth. 2000. *The Making of the GDR 1945–53. From Antifascism to Stalinism*. Manchester: Manchester University Press.

Reich, Ines and Kurt Finker. 1994. Reaktionäre oder Patrioten? Zur Historiographie und Widerstandsforschung in der GDR bis 1990. In *Der 20. Juli 1944. Bewertung und Rezeption des deutschen Widerstandes gegen das NS-Regime*, ed. Gerd R. Ueberschär, 126–142. Cologne: Bund-Verlag.

Rosen, Max. 2017. Section I.2: Share of World Population Living in Democracies, https://ourworldindata.org/democracy/#share-of-world-population-living-in-democracies Accessed 25 February 2017.

Rothfels, Hans. 1949. *Die Deutsche Opposition gegen Hitler*. Chicago: Regnery.

Rusinek, Bernd. 1994. Jugendwiderstand und Kriminalität. In *Der 20. Juli 1944. Bewertung und Rezeption des deutschen Widerstandes gegen das NS-Regime*, ed. Gerd R. Ueberschär, 291–308. Cologne: Bund-Verlag.

Satyanath, Shanker, Nico Voigtländer and Hans-Joachim Voth. 2017. Bowling for Fascism: Social Capital and the Rise of the Nazi Party. *Journal of Political Economy* 125 (2): 478–526.

Schlüter, Holger. 1995. *Die Urteilspraxis des nationalsozialistischen Volksgerichtshof.* Berlin: Duncker & Humblot.

Smith, Charles and Mark Farrales. 2010. Court Reform in Transitional States: Chile and the Philippines. *Journal of International Relations and Development* XIII: 163–193.

Solomon, Peter. 2007. Courts and Judges in Authoritarian Regimes. *World Politics* LX: 122–45.

Stadler, Karl. 1966. *Österreich 1938–1945 im Spiegel der NS-Akten. Sammlung Das Einsame Gewissen.* Vienna: Herold.

Steinbach, Peter. 1994. *Widerstand im Widerstreit: Der Widerstand gegen den Nationalsozialismus in der Erinnerung der Deutschen.* Munich: Ferdinand Schöningh.

Steinbach, Peter and Johannes Tuchel eds. 1998. *Lexikon des Deutschen Widerstandes.* Munich: Verlag C.H. Beck.

Steinberg, Hans-Josef. 1969. *Widerstand und Verfolgung in Essen 1933–1945.* Hannover: Verlag für Literatur und Zeitgeschehen.

Tate, Neal. 1995. Why the Expansion of Judicial Power? In *The Global Expansion of Judicial Power*, eds. Neal Tate and Torbjorn Vallinder, 27–37. New York: New York University Press.

The Economist Intelligence Unit. 2017. Democracy Index 2016: Revenge of the "Deplorables", January 25. http://www.eiu.com/public/thankyou_download.aspx?activity=download&campaignid=DemocracyIndex2016 Accessed on 25 February 2017.

Ueberschär, Gerd R. ed. 1994. *Der 20. Juli 1944. Bewertung und Rezeption des deutschen Widerstandes gegen das NS-Regime.* Cologne: Bund-Verlag.

Voigtländer, Nico and Hans-Joachim Voth. 2012. Persecution Perpetuated: The Medieval Origins of Anti-Semitic Violence in Nazi Germany. *Quarterly Journal of Economics* 127 (3): 1339–1392.

Voigtländer, Nico and Hans-Joachim Voth. 2015. Nazi Indoctrination and Anti-Semitic Beliefs in Germany. *Proceedings of the National Academy of Sciences* 112 (26): 7931–7936

Wagner, Walter. 1974. *Der Volksgerichtshof im nationalsozialistischen Staat.* Stuttgart: Deutsche Verlags-Anstalt.

Waldinger, Fabian. 2016. Bombs, Brains and Science: The Role of Human and Physical Capital for the Creation of Scientific Knowledge. *Review of Economics and Statistics* 98 (5): 811–831.

Wickert, Christl. 1994. Frauen im Hintergrund — das Beispiel von Kommunistinnen und Bibelforscherinnen. In *Das 'andere' Deutschland im Widerstand gegen den Nationalsozialismus: Beitrage zur politischen Überwindung der nationalsozialistischen Diktatur im Exil und im Dritten Reich*, 1st edition, eds. Helga Grebing and Christl Wickert, 200–25. Essen: Klartext.

Widner, Jennifer. 2001. *Building the Rule of Law.* New York: W.W. Norton.

Wippermann, Wolfgang. 1998. *Umstrittene Vergangenheit: Fakten und Kontroversen zum Nationalsozialismus.* Berlin: Elefanten Press.

Zank, Wolfgang. 1987. *Wirtschaft und Arbeit in Ostdeutschland 1945–1949: Probleme des Wiederaufbaus in der Sowjetischen Besatzungszone Deutschlands.* Munich: R. Oldenbourg Verlag.

Zarusky, Jürgen. 2011. Walter Wagners Volksgerichtshof-Studie von 1974 im Kontext der Forschungsentwicklung. In Walter Wagner, *Der Volksgerichtshof im nationalsozialistischen Staat mit einem Forschungsbericht für die Jahre 1974 bis 2010 von Jürgen Zarusky*, 993–1023. Munich: Oldenbourg Verlag.

Chapter 2
Sources

The collection of court records that forms the core of our research, Resistance as High Treason (*Widerstand als Hochverrat*), was the first of its kind to combine material stored in the archives of the Federal Republic of Germany and Austria with previously inaccessible files from the former German Democratic Republic, Czechoslovakia and Soviet Union (Zarusky and Mehringer 1998, p. 5). In the words of one of its coeditors, Jürgen Zarusky (Zarusky and Mehringer 1998, p. 13), *Widerstand als Hochverrat* is the "most comprehensive collection of sources in the history of inner opposition to National Socialism…the starting point of the edition is not the issue of a particular group or specific region but...rather the definition of political resistance that aimed to overthrow the regime, as defined by the regime in the form of high treason". The microfiche series it generated contains around 70,000 pages of court files that document the judicial prosecution of more than 6000 men and women charged with high treason and treason, and who appeared before the central courts of the Third Reich.

The overwhelming majority of these cases came before the People's Court (*Volksgerichtshof*), created in April 1934 to deal exclusively with those charged with high treason or treason. The selection of 1743 cases before the People's Court represents all cases of treason, high treason and conspiracy to commit high treason against German citizens where court case records have been located (in over 90% of all cases) (Zarusky and Mehringer 1998, pp. 5, 33–34).[1] This microfiche includes a wealth of information on the indictment, investigation, judgement and sentencing of each defendant. From these records, which appear in their original unanalysed form

[1] The number of cases and defendants who appear in the *Widerstand als Hochverrat* series are as follows (defendants in parenthesis): People's Court 1743 (5549); Supreme Court 80 (314) and Supreme Military Court 60 (168). During the Third Reich, cases of high treason were held before the following courts: People's Court (August 1934 to April 1945), Supreme Court or *Reichsgericht* (January 1933 to July 1934), Higher Regional Court or *Oberlandesgericht* (January 1933 to April 1934) and the Supreme Military Court or *Reichskriegsgericht* (1934–1945). After August 1934, most cases of high treason came directly under the jurisdiction of the People's Court, explaining the numerical preponderance of the People's Court.

W. Geerling, G. Magee, *Quantifying Resistance*, Studies in Economic History,
DOI 10.1007/978-981-10-6008-3_2

in the *Widerstand als Hochverrat* collection, we have been able to reconstruct systematically for the first time the profiles, family backgrounds and influences – and sentencing – of 4378 resisters charged with treason and high treason between 1933 and 1945. For each defendant, 259 variables have been collated, which together give rise to a database with 1,133,902 data points. Wherever possible, we have verified and supplemented our information with the entries for individual resisters found in the leading encyclopaedias of German resistance (Steinbach and Tuchel 1998; Steinbach and Tuchel 2004). Moreover, by examining the verdicts and sentencing of resisters, it also sheds a unique quantitative light on the deliberations of judges presiding over the People's Court and on how the Nazi justice system reacted to serious internal resistance from its citizens. Additional information on high-ranking Nazi-era judiciary has also been gleaned from the Federal Archive (*Bundesarchiv*) in Berlin, which holds the personal files (*Personalakten*) of judges who served on the Supreme Court and People's Court as well as the Nazi Party's membership records (*Mitgliedskarten*) which survived the war. We were able to locate personal files for 22 of the 38 presiding judges (21 of the 27 presiding judges from the People's Court)[2] and supplemented this knowledge with leading legal encyclopaedias of the Third Reich (Klee 2003). These files provided data on the judges' age, career prior to the Nazis coming to power (whether they had served on the bench during the Weimar Republic), whether they joined the Nazi Party (if so, when) and religious denomination.

Two key interconnected questions need to be addressed at this point. First, how representative of the most serious political crime in Nazi Germany is the database we have constructed? Second, which cases have we included in or excluded from it? As explained earlier, our database is ultimately framed by the source material contained in the *Widerstand als Hochverrat* collection. Only citizens of the old Reich, (pre-*Anschluß* border of Germany or *Altreich*), Austrians and Sudeten Germans (*Reichsangehöriger*) were included in its collection. According to Zarusky (Zarusky and Mehringer 1998, p. 13):

> …in a legal sense they were treated equally and the German and Austrian resistance operated under similar conditions, whereby most Austrian groups were united by the goal of overthrowing the National Socialist dictatorship with the restoration of an independent Austrian state. An important reason for the inclusion of cases against Austrians in the edition was to allow comparable investigations between the forms, groups and environments of resistance in Austria and the old Reich. The classification of nationality is orientated not on formal citizenship rather the context of the resistance activities. Therefore, a 'Germanised' former citizen of Poland who operated in the Polish resistance was not taken into consideration.

By definition, then, defendants who were not German citizens, such as Poles and Czechs, were excluded from the *Widerstand als Hochverrat* series (and by extension, our database). There were two exceptions to this rule. First, when foreigners

[2]The personal files can be found at *Bundesarchiv* BArch R 3001. The NSDAP membership cards can be located on microfilm at *Bundesarchiv* BArch NSDAP Zentralkartei 31 XX and *Bundesarchiv* BArch NSDAP Ortskartei 3200. Full details can be found in the reference list.

were involved in activities which took place in the context of German resistance – that is, the crimes were committed on German territory – they were often indicted alongside Germans. The Reichstag fire trial of December 1933 was the clearest example: four of the five defendants were foreigners, but the crime was attributed to the Communist Party of Germany (*Kommunistische Partei Deutschlands* or *KPD*), so all defendants appeared before the Leipzig Supreme Court. Second, many defendants who were involved in left-wing political cadre activities and fled abroad were later deprived of their German citizenship and so were technically stateless by the time their case was heard. Foreigners and stateless defendants together represent 4.5% of our database.[3]

The overwhelming majority of cases captured by the *Widerstand als Hochverrat* project were heard by the People's Court: 1743 cases (5549 defendants). A small number of cases – less than 8% – were tried before the Supreme Court and the Supreme Military Court. We decided to include cases from the Supreme Court, heard between 1933 and 1934, on the grounds that there was legal continuity between the Weimar and Nazi periods, and before the People's Court began hearing cases of high treason on 1 August 1934, the Supreme Court was the highest court in the land. In contrast, the Supreme Military Court operated with a different legal code to the German legal system and only heard cases involving military personnel, so its cases were excluded (Haase 1993; Gribbohm 2004).

A limitation of the sources we consulted was that some court records were missing or incomplete from the collection. There are a number of reasons for such gaps in the record. For a handful of cases, the system simply collapsed before they could be completed; for most of the others, their records were either lost or destroyed in the final stage of war. These incomplete cases have been removed from our database with one exception: those involving members of the 20 July conspiracy where the verdict survived but the indictment was missing. Given the high profile of these defendants, however, we were able to go back and piece together the missing information from various encyclopaedias and accounts of the 20 July conspiracy (Benz 2014; Steinbach and Tuchel 1998; Steinbach and Tuchel 1990). Taking into account the preceding limitations and exclusions, we nonetheless believe that our database is highly representative of the most serious political resistance in Nazi Germany. Expressed differently, its 1435 cases (4378 defendants) constitute nothing less than the entire population of Germans, Austrians and Sudeten Germans charged with treason and high treason and arraigned before the highest central courts of the Third Reich, for whom good records exist.[4]

[3] We used four citizenship categories: (i) citizens living within Germany's borders before the end of 1937 (*Altreich*), (ii) citizens incorporated within the German Reich after March 1938 (*Reichsdeutsche*), (iii) foreigner and (iv) stateless. The largest group was *Altreich*: 2764 (63.13%), followed by *Reichsdeutsche*: 1417 (32.36%), foreigner: 112 (2.56%) and stateless: 85 (1.94%).

[4] Our analysis includes 78.72% of all the cases in *Widerstand als Hochverrat* and 74.67% of defendants. Approximately 15–20% of cases had incomplete court files.

2.1 The Dataset

As mentioned in the previous section, the *Widerstand als Hochverrat* series contains a wealth of unprocessed information on the background, crimes and sentencing of defendants charged with treason and high treason between 1933 and 1945. Our research, through systematic analysis and the creation of a unique database, unlocks that information for the first time, affording in the process a rare quantitative overview of the most serious resistance to the Nazi regime. The richness of variables collected for each defendant is given in Appendix 2.1. These variables include a range of the personal characteristics and experiences of the defendants, the resistance groups they belonged to and other environmental and historical influences that shaped them, as well as data on the charges laid against them, the trial and the sentencing process they underwent, the verdicts they received and the judiciary who presided over their cases.

Of course, when reconstructing detailed profiles of resisters, we had to make a series of judgement calls as to how to classify certain aspects of their backgrounds, actions, affiliations and experiences. Some of these decisions perforce involved some degree of subjectivity: how does one consistently classify, inter alia, the level of education attained by defendants, the location of their resistance activities, the groups they belonged to, and even whether or not they defected during interrogation? In such instances, we based our judgements on the best information available to us in the historical record.

Take the example of educational attainment. The education system in Germany and Austria has several regional differences in terminology. We have adopted the following classificatory scheme:

(i) Tertiary (university): *Universität, Hochschule.*
(ii) High school advanced: *Gymnasium* which leads to an Abitur (final exam) and is the pathway to university.
(iii) High school medium: lower forms of secondary education including *Bürgerschule*, *Hauptschule*, *Mittelschule* and *Realschule.*
(iv) Primary school refers to the period of compulsory education: *Volksschule* or *Pflichtschule.*

Vocational/trade school with a focus on practical training is captured by a separate "vocation" variable. This would typically take the form of an apprenticeship (*Lehre* or *Lehrzeit*) or internship (*Praktikum*). In a typical blue-collar workshop, one would find the following hierarchy of workers: master (*Meister*), journeyman (*Geselle*), learned worker (*angelernter Arbeiter*) and apprentice (*Lehrling*).

The level of education and any vocational training usually appeared in the indictment following a short description of the alleged crimes, at the beginning of a section entitled: "personal background of the accused". Thus, for a 34-year-old baker from Breslau indicted in 1938, we are told: The accused visited the primary school in Breslau and Burkertswalde, then did a 4-year baker's apprenticeship and worked until August 1924, partly as a baker journeyman, also as a bread deliveryman

(Anklage 7 J 348/38, pp. 2–3). Another example describes the progression of the defendant from primary school through to obtaining a doctorate. The accused visited the primary and lower-level secondary school in Fürstenau and then a gymnasium in Duisburg. After taking his final exam, he studied philology and theology in Heidelberg, Bonn and Berlin, graduating with a PhD and successfully passing the state exam in 1902 (Anklage 11 J 29/44, pp. 2–3).

The records also provide insight into defendants' behaviour in custody. Prior to their appearance before the People's Court, each resister was investigated, arrested and interrogated. During interrogation, resisters were separated from their collaborators and encouraged to confess and assist the authorities in their investigations. Information obtained through these interrogations benefited the defendants who provided it and was also used by the authorities against their co-defendants. Through a careful reading of the individual case files, we are able to determine which resisters had defected during interrogation. This was possible because Gestapo internal reports, which were held in the unpublished case files, indictments and verdicts of the People's Court, would freely refer to a resister's "full and frank" confession or how the evidence provided by them was used to convict other resisters. One such report involved the interrogation of an 18-year-old juvenile from Vienna who was charged with writing anti-Nazi pamphlets and destroying an oil painting of Hitler. The report noted the defendant had made a full confession, and "he [now] feels shame for his actions and recognises his error". Furthermore, he wished to atone by writing a new work "in which he will show his change of heart toward National Socialism" (Anklage 7 J 104/44, p. 3; Urteil 5H 107/44, p. 4). Similarly, in another report about the interrogation of a resister from Leipzig, the prisoner, after much resistance finally "brought himself to a confession after being told that everyone had already spoken" (*Bundesarchiv*, VGH/Z H 286, p. 16). Failures to defect were also recorded by Gestapo interrogators and were typically attributed to the resisters' alleged obstinacy, remorselessness, immorality or lack of honour. In the Gestapo files of one Austrian woman, it was recorded that despite "persistent reminders to be honest" stretching over 4 months, she refused to cooperate with investigators (Anklage 8 J 377/42, pp. 4–6).

The *Erschliessungsband* to the *Widerstand als Hochverrat* microfiche series lists the location of resistance activities, as cited in the case files, by towns and cities (A-Z). Given the nature of the crimes involved, for example, the distribution of materials smuggled into or out of Germany, the work of émigrés abroad, a defendant's activities often took place in more than one location. As the territorial boundaries of Germany expanded to include Austria and the Sudetenland, then again during the war, so too did the location of resistance activities. As a result, it is neither practical nor feasible to rank locations of resistance activities by priority, as locations, like the participants and crimes, were not static. Their foci changed regularly across the period 1933–1945. Therefore, in line with the approach taken by the authors of *Widerstand als Hochverrat*, all locations where the resistance took place were recorded in our database.

Our database also captures some resistance activities undertaken in territories outside Germany. By outside territories, we are referring here to serious resistance by non-military citizens of the Greater German Reich, which took place outside the formal boundaries of Germany and Austria pre-*Anschluß*. Some of these territories were later incorporated into the Reich (the Polish corridor, Sudetenland), occupied (Baltic States, Belgium, Bulgaria, rest of Czechoslovakia, Denmark, France, Greece, Luxembourg, the Netherlands, Norway, rest of Poland, USSR and Yugoslavia) or held by fellow Axis partners (Hungary, Italy and Romania). In the period preceding the war, this resistance was predominantly the work of *KPD* and *SPD* cadre and émigrés operating on behalf of their respective leadership groups in exile. Activities were concentrated in neighbouring countries, usually confined to smuggling information, materials or people into or out of Germany and the restoration of illegal underground cells. In a typical case, a German émigré operated for the illegal *KPD* in Czechoslovakia, where he took part in communist training courses held between 1933 and early 1936. On 13 April 1936, he crossed the German border near Neugersdorf without a passport on behalf of the *KJVD*, carrying a sum of money, which he converted to 350 Reichsmarks without permission from a foreign currency agent. He transferred 300 Reichsmarks to a comrade in Berlin to be used for communist purposes (Anklage 16 J 287/36, pp. 2–3). Many émigrés were captured or repatriated to Germany after the annexation of the Sudetenland in October 1938 and the end of the Western campaign in June 1940. In one such case in the pre-war period, a *KPD* functionary from Saarland fled Germany in 1935. After spending a short period of time in an emigrant assembly camp in Alsace Lorraine, he resumed work for the *KPD* Central Committee, which at this time was in exile in Paris, delivering reports about atrocities from the Saarland and up-to-date information on the *Wehrmacht*. Furthermore, he smuggled communist pamphlets into the Saarland and represented the *KPD* in popular front and workers' committees with socialist and trade union representatives, between 1937 and 1939. The defendant was arrested by French authorities on 7 October 1941 and handed over to German police on 16 June 1942. By the time of the indictment, dated 4 January 1943, the defendant had been stripped of his German citizenship (Anklage 6 J 27/42 g, pp. 1–3).

In addition to determining the regional location of resistance, we have also assigned every city or town in Germany and Austria to an electoral district to enable a systematic analysis of the factors driving regional variations in arrest. For this purpose, we use the last truly democratic elections held before the dictatorships of Hitler and Dollfuß were established in November 1932 and November 1930, respectively.[5]

---

[5]The classification for Germany was problematic because there was a lack of continuity following the collapse of Imperial Germany (*Kaiserreich*). State and electoral boundaries overlapped. Free States were created, abolished or merged into other territories during the post-war period, and boundaries constantly changed. In the period 1924–1933, the Weimar Republic had 35 electoral districts (*Wahlkreise*) and 16 constituency associations (*Wahlkreisverbände*). We were able to get a breakdown of the vote for the main parties in each of the 35 electoral districts. In the case of Austria, we used data from the nine federal states (*Bundesländer*).

Resisters belonged to a vast array of political organisations and resistance groups. Not surprisingly, details of such political affiliations were clearly recorded in the files. Thus, in the period from June to November 1942, one case file informs us that Leopoldine Sicka, Anna Gräf, Franz Sikuta and Karl Mann were charged with belonging to an illegal political organisation, undermining the fighting spirit of the German armed forces and conspiring to commit arson and treason. Between 1941 and the beginning of 1942, the defendants in question had allegedly been active in the Communist Youth Association of Austria (*Kommunistische Jugendverband Österreichs* or *KJVÖ*) in the II. Region of Vienna, where Sicka, who had been a member of the Communist Party since 1939, operated as a regional leader (*Gebietsleiter*). Sikuta was an intermediary or contact (*Verbindungsmann*), while Gräf, who joined in the summer of 1940, and Mann were contacts at the district level (*Verbindungsmann im Gebietsmaßstab*). All defendants were involved in the printing and distribution in Vienna of subversive leaflets with defeatist messages, which were sent on to troops at the front. Mann also allegedly committed arson, while Sicka, Gräf and Sikuta were said to have conspired to commit arson (*DÖW* 3367, pp. 1–2).

To classify the allegiances of these and other resisters, we have adopted the typology used by Zarusky and Mehringer (1998, p. 56). For many organisations, there are self-chosen group designations and abbreviations, such as *KPD* and International Socialist Militant League (*Internationaler Sozialistischer Kampfbund* or *ISK*). In other cases, group designations were applied retrospectively because they referred to a particular event, such as the 20 July plot, or were introduced through the earliest resistance historiography (e.g. take the case of the Helmuth Hübener group: four juvenile resisters operating in Hamburg). Cross-linkage is common when using too narrow a "choice of terms" for group designations. Therefore, broader collective names were often used for some resistance groups. For example, the 20 July group is used to capture all those connected with the conspiracy to kill Hitler, or Buendische Youth (*Bündische Jugend*) for youth who belonged to different associations, including hiking groups and boy scouts. In a few cases, classification is based not on resistance groups in the narrow sense but organisations, in whose realm the defendants operated, such as the National Committee for a Free Germany (*Nationalkomitee freies Deutschland* or *NKFD*).

Finally, actions carried out by individuals, either working alone or with other non-aligned individuals, were classified as "not organised". Such unorganised resistance became more common as the war deteriorated. It was often situational in nature, occurring while the defendant was drunk, after a dispute, in protest at a legitimate grievance, or during an air raid. Thus, one such resister, a 49-year-old grinder from Berlin, was arrested after writing the words "mass murderer" on a picture of Hitler in the common room of an air raid shelter in Berlin in April 1942 (Anklage 10 J 159/42, p. 1).

Appendix 2.1: Variables Used in the Quantifying Resistance Database

Personal and Familial Characteristics and Experiences

Citizenship

Altreich: citizens living within Germany's borders before the end of 1937
Foreigner
Reichsdeutsche: citizens incorporated within the German Reich after March 1938
Stateless

Criminal History

Previous Convictions
Previous Convictions: Malicious Gossip
Previous Major Political Convictions: High Treason, Treason
Previous Minor Political Convictions: Defamation
Previous Minor Political Convictions: Illegal Meetings and/or Writings
Previous Minor Political Convictions: Public Disorder
Previous Minor Political Convictions: Resisting State Authorities
Previous Non-Political Convictions: Other Crimes
Previous Non-Political Convictions: Property Crimes
Previous Non-Political Convictions: Sexual Offences
Previous Non-Political Convictions: Violent Crimes
Previous terms of imprisonment

Education (Highest Educational Achievement)

High School Advanced
High School Medium
Primary
Tertiary

Marital and Familial Status

Divorced
Engaged or married (de jure and de facto)
Has children: number
Widowed

Occupation

Blue collar
Business/property owner
Employed
High ranking state official

Main occupation
Military
No particular occupation
Not in the labour force
Political cadre
Professionals
Secondary occupation
Senior military
Service sector
Student
Unemployed
Vocational training
White collar

Personal

Age at indictment
Came of age during the 1918–1919 revolution
Came of age during the period of hyperinflation
Chronic health problems
Date of birth
Denied entry or expelled from Hitler Youth or League of German Girls
Émigré
Family member killed during the war
Gender
Immediate family engaged in anti-regime activities
Mischlinge: defendant had partial Jewish ancestry
Parents alive
Place of birth
Place of residence
Served in WWI
Served in WWII
Spent time outside the Reich in a private capacity

Characteristics of Charges and Trial

Charges

Major High Treason: Crimes Abroad
Major High Treason: Foreign Broadcasts
Major High Treason: Illegal Organisation
Major High Treason: Illegal Writings

Major High Treason: Separatism
Major High Treason: Subversion of Police or Armed Forces
Major High Treason: Undermining the Fighting Spirit of the German Armed Forces
Major High Treason: Violence Against Hitler
Minor High Treason: Aiding and Abetting
Minor High Treason: Conspiracy to Commit High Treason
Minor High Treason: Failure to Report
Treason

General

Case Number
Co-defendant/s cooperated with the authorities during the investigation and/or trial
Date of arrest
Date of indictment
Defendant cooperated with the authorities during the investigation and/or trial
Group resistance
Juvenile co-defendants
Juvenile defendants
Number of defendants
Senate number
Violent resistance

Location of Crime (*Altreich*)

Baden-Wuerttemberg
Bavaria North West
Bavaria South East
Brandenburg (Berlin)
Brandenburg II
Crime committed in a German region
East Prussia
Hessen
Lower Saxony and Bremen
Pomerania-Mecklenburg
Rhineland North
Rhineland South
Saarland
Saxony
Saxony-Anhalt and Thuringia
Schleswig-Holstein and Hamburg
Silesia
Westphalia

Location of Crime (Austria)

Burgenland
Carinthia

Lower Austria
Salzburg
Styria
Tyrol
Upper Austria
Vienna
Vorarlberg

Location of Crime (Outside Territories)

Belgium
Czechoslovakia
Denmark
France
Luxembourg
Netherlands
Norway
Poland
South and South Eastern Europe: Bulgaria, Greece, Hungary, Italy, Romania and Yugoslavia
USSR: includes Baltic States

Verdict and Sentencing

Convictions

Date of sentencing
Major High Treason: Crimes Abroad
Major High Treason: Foreign Broadcasts
Major High Treason: Illegal Organisation
Major High Treason: Illegal Writings
Major High Treason: Separatism
Major High Treason: Subversion of Police or Armed Forces
Major High Treason: Undermining the Fighting Spirit of the German Armed Forces
Major High Treason: Violence Against Hitler
Minor High Treason: Aiding and Abetting
Minor High Treason: Conspiracy to Commit High Treason
Minor High Treason: Failure to Report
Treason
Violent Crime

President of People's Court

Bruner
Freisler

Haffner
Rehn
Thierack

Presiding Judge

Albrecht
Bruner
Bünger
Coenders
Coninx
Crohne
Diescher
Diester
Doorberg
Driver
Engert
Fikeis
Freisler
Granzow
Greulich
Großpietsch
Hartmann
Illner
Jenne
Köhler
Lämmle
Linz
Löhmann
Makart
Mengelkoch
Merten
Müller
Niethammer
Pietsch
Presiding Judge a member of the Nazi Party
Presiding Judge an old party comrade (*Alte Kämpfer*)
Preussner
Rehn
Rheinisch
Schaad
Schauwecker
Springmann
Stier
Thierack
Zieger

Sanction

Acquitted/dismissed
Death
Penitentiary
Prison
Term of imprisonment

Resistance Groups

Austrian

Across Party Lines
Armed Resistance/Partisans
Austrian Communist Help Organisation
Austrian Trotskyists
Catholic/Conservative/Legitimists
Communist Party of Austria
Other Socialist
Revolutionary Socialists of Austria
Social Democratic Party of Austria
Viennese Mischlingsliga
Youth

German

Anarchist
Black Front
Buendische Youth
Catholic
Communist Help
Communist Party of Germany
Communist Party of Germany (Opposition)
Émigré
Leipzig Meuten
National Conservative
Other Communist Splinter
Other Socialist
Separatist
Social Democratic Party of Germany
Socialist Workers Party of Germany
Trade Union
Trotsky
White Rose
Youth
20 July

Environmental Factors

Membership of Organisations (Austrian)

Austrian Nazi
Christian Social
Communist
Fatherland Front
Groups dedicated to protecting the Austrian Republic
Home Guard
Legitimist
National Conservative
Pan-German Nationalist
Social Democrat

Membership of Organisations (German)

Anarchist
Black Front
Centre Party
Communist
Free Corps or Association of Front Soldiers
German Nazi
Groups dedicated to protecting the Weimar Republic
Hiking
National Conservative
Non-Political and Non-Religious
SA or SS
Social Democrat
Socialist
Trade Union
Trotskyist
Youth Street Gang

Religion

Catholic
Evangelical
Mormon
Religious Convictions

Historical Influences

Arrest Date

Weimar Period (up to 4 March 1933)
5 March 1933 to 14 July 1933
15 July 1933 to 2 May 1934
1 September 1939 to 8 May 1945
Barbarossa to Stalingrad (22 June 1941 to 2 February 1943)
Stalingrad to D-Day (3 February 1943 to 6 June 1944)
Post D-Day (7 June 1944–)

Sentencing Date

Weimar Period (up to 4 March 1933)
5 March 1933 to 14 July 1933
15 July 1933 to 2 May 1934
1 September 1939 to 8 May 1945
Barbarossa to Stalingrad (22 June 1941 to 2 February 1943)
Stalingrad to D-Day (3 February 1943 to 6 June 1944)
Post D-Day (7 June 1944–)

References

Primary Sources

(A) Archive Files

(i) Widerstand als Hochverrat

Anklage 16J 287/36 In: *Nationalsozialismus, Holocaust, Widerstand und Exil 1933–1945*. Online-Datenbank. De Gruyter. Accessed Jul 22, 2015. http://db.saur.de/DGO/basicFullCitationView.jsf?documentId=wh259

Anklage 7J 348/38 In: *Nationalsozialismus, Holocaust, Widerstand und Exil 1933–1945*. Online-Datenbank. De Gruyter. Accessed May 24, 2013. http://db.saur.de/DGO/basicFullCitation-View.jsf?documentId=wh476

Anklage 6J 27/42g In: *Nationalsozialismus, Holocaust, Widerstand und Exil 1933–1945*. Online-Datenbank. De Gruyter. Accessed Jul 22, 2015. http://db.saur.de/DGO/basicFullCitatiwon-View.jsf?documentId=wh1154

Anklage 8J 377/42 In: *Nationalsozialismus, Holocaust, Widerstand und Exil 1933–1945*. Online-Datenbank. De Gruyter. Accessed Feb 16, 2014. http://db.saur.de/DGO/basicFullCitationView.jsf?documentId=wh2148

Anklage 10J 159/42 In: *Nationalsozialismus, Holocaust, Widerstand und Exil 1933–1945*. Online-Datenbank. De Gruyter. Accessed Oct 11, 2013. http://db.saur.de/DGO/basicFullCitationView.jsf?documentId=wh2495

Anklage 7J 104/44 In: *Nationalsozialismus, Holocaust, Widerstand und Exil 1933–1945*. Online-Datenbank. De Gruyter. Accessed Feb 12, 2014. http://db.saur.de/DGO/basicFullCitationView.jsf?documentId=wh1980

Anklage 11J 29/44 In: *Nationalsozialismus, Holocaust, Widerstand und Exil 1933–1945*. Online-Datenbank. De Gruyter. Accessed Oct 7, 2013. http://db.saur.de/DGO/basicFullCitationView.jsf?documentId=wh711

Urteil 5H 107/44 In: *Nationalsozialismus, Holocaust, Widerstand und Exil 1933–1945*. Online-Datenbank. De Gruyter. Accessed Feb 12, 2014. http://db.saur.de/DGO/basicFullCitationView.jsf?documentId=wh1981

Zarusky, Jürgen and Hartmut Mehringer eds. 1998. *Widerstand als Hochverrat 1933–45. Die Verfahren gegen deutsche Reichsangehörige vor dem Reichsgericht, dem Volksgericht und dem Reichskriegsgericht*. Munich: K G Saur. Microfiche edition.

(ii) Bundesarchiv

Bundesarchiv, Case Files of the *Volksgerichtshof*, VGH/Z H 286.

Reichsministerium der Justiz, *Personalakten*, R 3001

The surname of the judge and signature number in parenthesis: Albrecht [50165, 50167], Bruner [52916], Coninx [53671], Crohne [53759], Diescher [54191], Diester [54196], Engert [55261, 55265], Freisler [56247], Granzow [57872], Greulich [57983], Grosspietch [58154], Hartmann [59107, 59108], Illner [61600], Köhler [63914], Lämmle [65592, 65593], Löhmann [66929], Merten [68165, 68166], Schaad [73551], Springmann [76972], Stier [77504], Thierack [78253], Zieger [81308, 81309].

The NSDAP membership cards can be located on microfilm at *Bundesarchiv BArch NSDAP Zentralkartei* 31 XX and *Bundesarchiv BArch NSDAP Ortskartei* 3200. The surname of the judge and signature number in parenthesis: Diescher [ZK 31XX E0111], Diester [ZK 31XX E0112], Freisler [OK 3200 E0079], Illner [OK 3200 J0013], Köhler [OK 3200 J0013], Merten [OK 3200 O0067], Schaad [OK 3200 T0003], Zieger [ZK 31XX T9157].

(iii) Dokumentationsarchiv des österreichischen Widerstandes

DÖW 3367

Secondary Sources

Benz, Wolfgang. 2014. *Der deutsche Widerstand gegen Hitler*. Munich: C.H. Beck.

Gribbohm, Günter. 2004. *Das Reichskriegsgericht: Die Institution und ihre rechtliche Bewertung*. Berlin: Berliner Wissenschafts Verlag.

Haase, Norbert. 1993. *Das Reichskriegsgericht und der Widerstand gegen die nationalsozialistische Herrschaft: Katalog zur Sonderausstellung der Gedenkstätte Deutscher Widerstand in Zusammenarbeit mit der Neuen Richtervereinigung*. Berlin: Gedenkstätte Deutscher Widerstand.

Klee, Ernst. 2003. *Das Personenlexikon zum Dritten Reich*. Frankfurt am Main: Fischer-Taschenbuch-Verlag.

Steinbach, Peter and Johannes Tuchel eds. 1990. *Gedenkstätte Deutscher Widerstand. Katalog zur Sonderausstellung der Gedenkstätte Deutscher Widerstand in Zusammenarbeit mit der Neuen Richtervereinigung*. Berlin: Gedenkstätte Deutscher Widerstand.

Steinbach, Peter and Johannes Tuchel eds. 1998. *Lexikon des Deutschen Widerstandes*. Munich: C.H. Beck.

Steinbach, Peter and Johannes Tuchel eds. 2004. *Widerstand gegen die nationalsozialistische Diktatur 1933–1945*. Bonn: Lukas-Verlag.

Chapter 3
Times and Places

In this chapter our analysis of serious internal resistance to Hitler's regime begins with an examination of the overall structure and composition of that resistance from the inception of Nazi rule in January 1933 through to the collapse of the German legal system in early 1945. The chapter seeks to answer two fundamental questions in particular, one temporal and one geographic: namely, how did levels of serious resistance change across time and what regions were most active in this resistance? Along the way, a number of secondary issues are also considered: inter alia, can a distinct periodisation of resistance be identified, how did the different phases of the war impact resistance activities, did the intensity of such activities in any location merely reflect its population size, and is it more appropriate to talk of the commonality or distinctiveness of regional resistance experiences? The answers to such questions gleaned from our quantitative analyses of the available data on treason and high treason, we believe, shed new light on, and challenge aspects of, our understanding of the path of serious resistance within Nazi Germany.

3.1 Periodisation

A common feature of much of the historiography of German resistance is its focus on discrete, often pivotal, events—and the groups that arose in response to them—in the struggle against the National Socialist state. Here, one thinks most typically of the Battle of Stalingrad, the landings at Normandy, the White Rose and the bombing plot configured around Claus Schenk Graf von Stauffenberg. The history of resistance is frequently told as a narrative leading up to, and in the aftermath of, these climacteric moments. The periodisation of German resistance is thus often shaped, both consciously and unconsciously, around these highly visible actions or

W. Geerling, G. Magee, *Quantifying Resistance*, Studies in Economic History,
DOI 10.1007/978-981-10-6008-3_3

conspiracies.[1] Rarely, if at all, is it driven by a systematic analysis of all serious resistance activities, including, importantly, those very many acts, which have escaped publicity and attention and been consigned to the penumbra of historical discourse. An advantage of quantitative analysis, when properly constructed, is that it restores these acts of resistance into the broader narrative.

To date, the most systematic attempts to capture the trends and periodisation of serious resistance have been provided by legal historians, who have constructed annual series of the number of proceedings and defendants charged with high treason and treason and tried before the People's Court from official publications (Schlüter 1995; Marxen 1994; Lauf 1994). There are some variations in these series between researchers, but, as they all draw on the same source material, the differences occur in only a few years at the beginning and the end of the court, when the records are less clear or complete. The broad story told by each of these series is virtually identical. For the sake of conciseness, Fig. 3.1 presents just the series for proceedings and sentenced defendants reported by Holger Schlüter, the latest and arguably most authoritative account yet produced.

These series, in particular the number of sentenced defendants, outline the usual periodisation of resistance. It is an easily told story. The resistance experienced by the regime at its inception is decisively broken by the mid-1930s, and the number of convictions goes into decline for a number of years thereafter.[2] A mild upswing in proceedings and sentencing follows in the wake of the annexations of Austria and Czechoslovakia and then the military successes of the first stage of the war, which together had the effect of bringing many exiled resisters back within the reach of the regime. From 1941, both series steadily pick up from a fairly low base, gathering steam, especially after Stalingrad and as the war deteriorates. They finally reach their zenith in 1944, the year of greatest resistance activity. As Richard Evans (2008, p. 625) succinctly explains:

> the military catastrophes of the spring and early summer of 1944 led to an upsurge of resistance not only in occupied Europe but also in the German Reich itself. Already, the defeats of the previous year had spread disillusion with the regime. The devastating effects of the bombing weakened the authority of the regime still more. Nevertheless, open acts of resistance or defiance were still rare.

By the early months of 1945, the system, including its courts, was in the process of collapsing. Figure 3.1, thus, appears at first blush to confirm quantitatively the vital importance of Stalingrad, D-Day and the 20 July 1944 plot in the advent of serious

[1] One such book is Joachim Fest's *Staatsstreich: Der lange Weg zum 20. Juli* (1997), in which he challenges the notion of everyday resistance and focuses on the men and women who had the ability to challenge the regime effectively. For a critical assessment of the emphasis on key events and national conservatives involved in the 20 July plot, see Kershaw (2015).

[2] Thus, as Benz (2014, p.11), for example, has argued: "the early resistance of the working class was already pulverised by the time the bourgeois elites came into opposition against the regime. Again, it took time until the military elite, civil servants and diplomats resolved to overthrow the dictatorship and plan a new order".

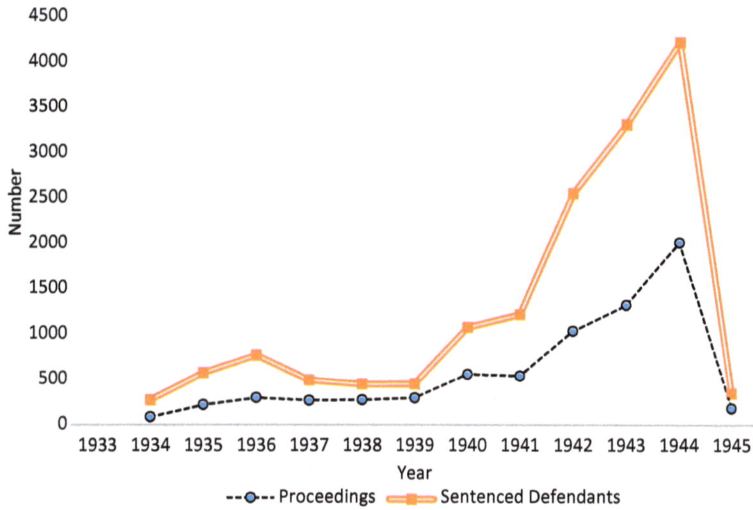

Fig. 3.1 Proceedings and sentenced defendants in the People's Court (Schlüter 1995, p. 37)

resistance in the German Reich. The figure, however, also reveals a dimension of the story less commonly recognised in the broader literature: that serious resistance activity neither disappeared nor was in permanent free fall between the defeat of all major organised Weimar-era political opposition by the mid-1930s and military disaster at Stalingrad.[3] In fact, a considerable amount of activity continued in this period, a fact that we will look more closely at later in the chapter.

But first, a major limitation of using series such as those in Fig. 3.1 as the basis for a periodisation of serious resistance is worth considering: the temporal separation of what is being captured in the series from the actual acts of resistance themselves. The number of proceedings opened and defendants sentenced are legal processes that occur at the very end of the process that began with the thought of resistance and became real with its consummation or planned consummation. The time gap between the act or intention to resist and the beginnings of the legal process, let alone the sentencing of a defendant, could be considerable. Moreover, the duration of that gap could vary according to how each different branch of the

[3] Interestingly, the annual number of people interrogated at the Gestapo's House Prison on Prince Albrecht Street in Berlin also reveals this pickup in activity, especially between 1938 and 1940, a year in which the Gestapo were interrogating more people than ever before. Interrogations, of course, often do not lead to charges, so they are not necessarily a good match for actual resistance levels. But the timing of the increase in interrogations is consistent with the uplift in cases seen in Fig. 3.1. Unfortunately, due to missing records and interrogations being shifted to other locations with the onset of Allied bombing, the trend in interrogations found in the House Prison data cannot be followed beyond 1940; see Topography of Terror Foundation (2014, pp. 214–35).

authorities undertook their tasks. Some cases of resistance, for example, were easier to detect and investigate than others; the time it took the state's legal bureaucracy to process a case depended on the legislation and regulations it operated within, as well as on the workload it was experiencing. Some cases, when they came to court, took longer to play out than others. The disruptions caused by Allied bombing and the intensifying war effort also impinged. The key point here is that sentencing or the beginnings of proceedings against a defendant could occur a long time after the act of resistance itself. Judicial records indicate that the gap between arrest and sentencing showed remarkable stability, however, ranging typically between 13 and 16 months. The mean gap was in fact 431 days, or just over 14.5 months. For most of the regime's existence, it was the time between arrest and indictment that explained most of the gap (about 75% of it, suggesting the importance of investigation and establishing formal charges in determining processing times). The exceptions were the years 1938 and 1939, (and less so in 1943–1944) when the period from indictment to sentencing (the judicial part of the process) accounted for between 31 and 45% of the gap.[4]

Closer to the act of resistance or at least the time when it registers with the authorities, than stages in the defendant's trial, is the arrest of the defendant on charges of high treason or treason.[5] Figure 3.2 presents the annual number of such arrests, which eventually ended up in trials before the Supreme or People's Court. It tells a different story from that depicted in Fig. 3.1. The early stage of the regime is characterised by two great waves of arrests in 1933 and 1935 rather than steady growth until 1936. The period between 1936 and 1940 also reveals an amazing stability in arrests (at around 200 a year). There is, then, a dramatic pickup in arrests in 1941 and 1942, which then start to drop off in 1943, a trend which continues into 1944. As the People's Court petered out in the first quarter of 1945, little can be read into the 1945 figure. One of the key differences uncovered by looking at arrest rather than trial-related dates is the pronounced increase in serious resistance activ-

[4]The average time taken to proceed from arrest to trial and sentencing has been calculated from the dates provided in the case records of each defendant. Patterns in the 'processing' time of defendants are discussed further in Chap. 6.

[5]In principle, of course, the best indicator for the correct periodisation of resistance activities is, in fact, the date on which the acts occurred. In practice, however, there are a number of issues in pursuing such an approach. First, the beginning date of a defendant's activities is often not known. Second, in the majority of cases, resistance was not a single discrete act but rather one that may have occurred over extended or indeed intermittent periods. Third, the gravity of resisters' activities may have changed over time, too. It was not uncommon for a resister to have come to the attention of the Gestapo at a fairly early date, just simply as a result of their being a possible associate of a known resister. He or she may not actually have engaged in any illegal activity themselves at that point of time. Indeed, as was often the case, they may have been drawn into the world of serious resistance in a rather piecemeal manner, starting out on relatively minor acts of disobedience, only crossing the line into high treason or treason from the regime's perspective a number of years after surveillance had begun. The key point is that, given the information available in the records, determining precisely when serious resistance began *in a consistent manner* is deeply fraught with difficulties. Date of arrest on charges of treason or high treason appears the event closest to the act of resistance that can be consistently captured.

Fig. 3.2 Arrests for treason and high treason, 1933–1945

ity it reveals *prior* to the defeat at Stalingrad. Moreover, this growth of resistance cannot be seen as part of a steady progression upwards as the war advanced. Rather, it peaked in 1942 and slowly declined thereafter. The key year of serious resistance was, in fact, 1942, not 1944. That is, resistance, as gauged by arrests, was at its most active before both the demise of the Sixth Army at Stalingrad and the opening up of the second front in the West. This finding suggests that the invasion of the USSR and the military setbacks of late 1941 and early 1942 may have played a much more important role in galvanising resistance than previously thought.[6] Official figures for the number of individuals held on remand by the People's Court, plotted in Fig. 3.3, provide further support for the contention that the strongest surge of activity from the regime's perspective occurred in 1941 and 1942 rather than later. Indeed, the years that followed also witnessed a rapid diminution of the numbers of people being taken into remand.

Figure 3.2 also identifies, what at first appearance, seems a curious stability of arrest numbers between 1936 and 1940. One explanation could be that in this period after the defeat of large-scale organised resistance, the Gestapo, like the NKVD (Soviet Secret Police) in Stalinist Russia, may have operated according to predetermined annual arrest targets. This, however, does not appear likely. The literature on

[6] Evans (2008, p. 625) hints at such a pickup in activity in 1942: "it was only in 1942, after the defeat of the German army before Moscow, that clandestine Communist resistance groups began to emerge again, in strongholds of the industrial working class like Saxony, Thuringia, Berlin and the Ruhr." However, he does not detect the scale of the increase taking place and continues to believe that 1944 was the year of peak activity.

Fig. 3.3 Individuals held on remand by the People's Court, 1934–1944 (Wagner 1974, pp. 877–78) (*Note*: The figures for the years 1935 through 1942 were taken directly from those reported in Wagner's Table 5(a) found on p. 878. According to Wagner (p. 878), the figures for 1943 and 1944 reported in the official publication from which he constructed his tables accidentally confused the number of new people on remand in each half year (except for the second half of 1943) with the total number of those on remand. On examination, Wagner's assessment appears sound and the figures used above take this correction into consideration. Wagner also claims that the total number of people reported in the official publication for the second half of 1943 mistakenly adds a '1' at the beginning of the provided number, viz. it reads as 16,671 instead of 6671. There is no evidence to support this contention, however. Indeed, as Fig. 3.3 illustrates, that official number is very much in line with the trend of people held on remand in that period)

the German secret services provides no evidence of the existence of such quotas or targets; indeed, one historian (Moorhouse 2010, p. 222) explicitly states that the Gestapo did not have targets.[7] Data on interrogations at the Gestapo's House Prison

[7]Another researcher, McDonough (2015, p. 9), emphasises 'the high level of autonomy the Gestapo was given to deal with cases, and the often exhaustive amount of time it devoted to them'. One of the potential limitations of using arrest data is that the timing of arrest may in some instances have been chosen by the Gestapo for strategic purposes. As Evans (2008, p. 625) has noted, 'as a precaution, too, the Gestapo arrested and incarcerated a number of former Communist functionaries on the invasion of the Soviet Union in case they should start a campaign of subversion'. What impact did such precautionary actions have on arrests for treason and high treason? Let's consider the instance mentioned by Evans. In the first 3 months of the invasion of Russia, June to August 1941, 54 arrests were, on average, made per month. The average monthly rate for 1941 as a whole—and for each half of that year—was 35. These figures, therefore, indicate that there was a pickup in arrests at this time, some of which may have been due to the factor raised by Evans. However, it is important to note that the numbers being arrested on average in that 3 month period were not that unusual. More than 50 arrests a month had previously occurred in July 1940 and January 1941, as well as on other occasions in the pre-war era. Moreover, the scale of the effect could not have been that large. At most, 'strategic arrests' could have added an extra 23 arrests to the average monthly arrest total at the beginning of Operation Barbarossa. In reality, it was almost certainly a lot less

in Berlin in period, moreover, provides no evidence of quantitative targets: there were wide variations on the numbers interrogated between years (Topography of Terror 2014).[8] On closer inspection of the data, one finds that the real reason for the stability of arrest numbers between 1936 and 1940 lays in the interactions of different trends occurring in various parts of the Third Reich and its occupied territories across this period.

Figure 3.4 depicts the annual number of arrests for treasonous or high treasonous activities in Germany as it existed before the end of 1937 (*Altreich*), Austria and by Germans and Austrians outside the Greater German Reich (i.e. in occupied territories). Three different paths emerge. Arrests in Germany declined steadily from 200 in 1936 through to 64 in 1940. There was a modest increase in arrests in 1941, followed by a steep continuous rise from 1942 that reached its peak in 1944. As for Austria, up to 1937 resistance activities there resulted in relatively small numbers of arrests. Following the annexation of Austria (*Anschluß*), not surprisingly, these numbers grew year on year until 1942. The most rapid growth of arrests occurred in 1941 following the start of Operation Barbarossa and in 1942, after which the numbers began to subside steadily, so much so that by 1944, the number of arrests had declined to their lowest level since 1940.[9]

Arrests for activities outside the Reich occurred throughout the entire Nazi period and in many instances resulted from the regime catching up with resisters who had previously fled. Much of this resistance activity was planned and led by Germans and Austrians who had sought refuge in lands then beyond Nazi control: most notably, France, Holland, Czechoslovakia, Denmark and the USSR (Union of Soviet Socialist Republics). The surge of arrests of 1935 contained a significant component of such outside resistance activities. As these territories beyond the Greater German Reich were steadily annexed or brought under the control of the Third Reich between 1938 and 1941, many of these exiles were either captured or handed over by local authorities to the Nazis, arrested and put on trial.[10] From 1942, as the expansion of German control first faltered and then began contracting, the arrests of Germans and Austrians for activities outside the Reich began to dwindle to significantly lower levels.

---

than that, as there is very good reason to believe that communist activity (and arrests) would have picked up after the invasion irrespective of what the Gestapo had done. It is also worth noting that from September to December 1941, the monthly rate fell back to 22, a finding which suggests that the mid-1941 increase in arrests was short lived. In short, the data available to us indicate that the Gestapo's strategic considerations were not the driving forces behind the arrest series at this time.

[8] It is also worth pointing out that there is no significant statistical correlation between the volume of Gestapo interrogations in Berlin and the number of arrests for treason or high treason. This finding suggests that interrogation numbers are not good predictors of arrests and indictments for serious resistance.

[9] The number of daily reports of potentially illegal political activity recorded by the Viennese Gestapo broadly follows the trends exhibited in Fig. 3.4 for Austria. It peaks in 1941 and falls away afterwards, especially from 1943; see Bailer and Ungar (2013, p. 171).

[10] For a short overview of resistance in exile, see Benz (2014, pp. 92–102).

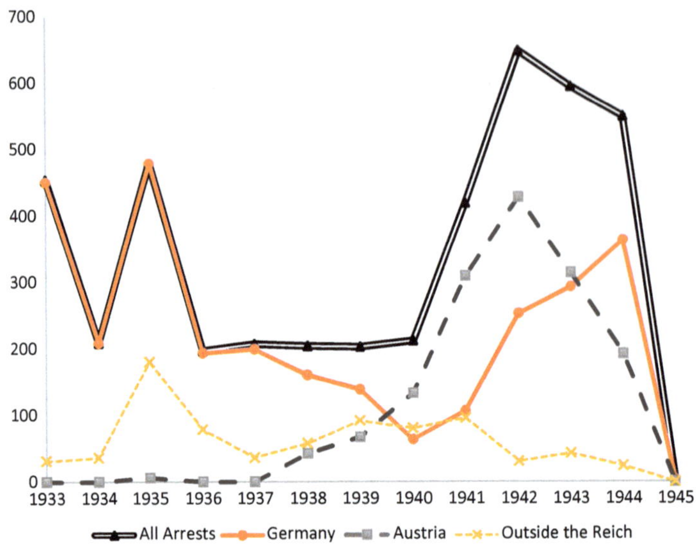

Fig. 3.4 Incidence of treason and high treason: Germany, Austria and outside the Reich (*Note*: All arrests sum to 4378 (1 per defendant)). The German, Austrian and outside the Reich breakdowns reflect incidence, i.e. where the activity/activities took place. These numbers sum to more than 4378 because many resisters operated in multiple locations: Germany (2916), Austria (1491) and outside the Reich (791))

Three further implications can be drawn from Fig. 3.4. First, it is meaningful to talk about multiple and distinct resistance experiences in the Third Reich. This was not just because organisationally, as Ernst Hanisch (1987, pp. 163–176) has pointed out, there was a 'near full separation…between Austrian and German resistance groups', but also because the stories of resistance in Germany, Austria and beyond were the same in neither pattern nor scale. Moreover, it is worth pointing out that arrests for treasonous and high treasonous activities based in Germany did not always dominant quantitatively. Indeed, from 1940 through to the end of 1943, more arrests occurred for activities outside Germany than inside.

Second, following from the previous implication, it is evident that the stability of arrests for treason and high treason between 1936 and 1940 seen in Fig. 3.2 arose from the fact that the decline in arrests in Germany during this period—which is the aspect of this period that gets most frequently reported in the literature—was matched by increases in arrests for resistance activities elsewhere following the annexation and occupations of Austria, Czechoslovakia, France and the low countries.

Third, the view that it was the perception of imminent defeat of the regime, first made abundantly clear in 1944 by the success of the Normandy landings, which drove the biggest upswing in resistance activity, resonates only with the German

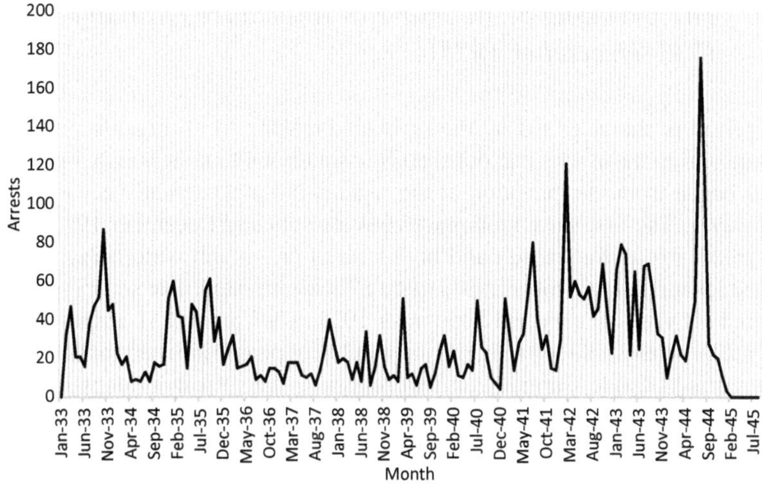

Fig. 3.5 Monthly arrests for treason and high treason, January 1933–April 1945

experience. In Austria and outside the Reich—and indeed overall—resistance activities actually peaked significantly earlier in 1941 and 1942.

Figure 3.5, which gives monthly arrest figures for treason and high treason throughout the Third Reich, provides even further granularity on the periodisation of arrests. By combining the information contained in Figs. 3.4 and 3.5, five distinct periods of serious internal resistance to the National Socialist state emerge from the data.[11]

[11] We acknowledge that the periodisation we offer here differs significantly from those proposed by others. Marxen (1994, p. 34), for example, suggested a three-part periodisation: (i) pre-war, (ii) main phase of the war (September 1939 to February 1944) and (iii) end phase of the war (March 1944 to April 1945). Schlüter (1995, p. 142), by contrast, puts forward four distinct phases in his analysis of oppositional cases before the People's Court: (i) pre-war, (ii) September 1939 to October 1941, (iii) November 1941 to February 1944 and (iv) March 1944 to December 1944. In accounting for the differences between these periodisations, it is worth noting that both of these legal historians chose their phases fairly arbitrarily and as part of an analysis of the sentencing patterns of the People's Court, not resistance per se. Their phases, thus, are not based on a reading of the data with respect to arrests, nor for that matter does the definition of 'oppositional' cases used by Schlüter align with that employed in this book or used by the broader resistance literature. It is also noteworthy that both Marxen and Schlüter's periodisations separate the resistance of the last year of the war from the preceding three war years. No clear explanation for doing so is provided. It may be speculated that one reason may be the untested assumption of the authors that the highly visible conspiracies of the final year were somehow distinctive from what had transpired before.

3.1.1 The Breaking of Weimar-Era Resistance (January 1933–December 1935)

This period is characterised by the regime's breaking of all organised large-scale resistance from those who had opposed Nazism throughout the Weimar period. The arrests began from the inception of the regime, but most actually occurred in two great waves. The first wave took place between July and December 1933, when on average 53 arrests for treason and high treason per month were being made, and reached its zenith in October 1933, when 87 were arrested. The second great wave occurred between January and September 1935, when around 61 arrests per month were made. These included a significant number of people arrested for activities that had occurred outside Germany, in part reflecting the shift in strategy of the main oppositional parties, the Social Democratic Party of Germany (*Sozialdemokratische Partei Deutschlands* or *SPD*) and Communist Party of Germany (*Kommunistische Partei Deutschlands* or *KPD*), towards the continuation of resistance activities from the relative safety of exile.

3.1.2 Relative Stability and the Emergence of New Sources of Resistance (January 1936–December 1940)

Following the defeat of the main oppositional groups in Germany, a period of relative stability from the regime's perspective set in. Arrests fell and remained for the most part at around 17 per month, a stability broken only by short-lived spikes of fairly moderate amplitude in part associated with the *Anschluß* and acquisition of occupied territories. The most pronounced spike took place in March 1939 when 51 arrests in total were made. Activities outside Germany played an increasingly important role in this period. The rise of Austrian and outside incidences at this time compensated for the decline in serious German-based resistance. From early 1940, there were more Austrians being arrested for treason and high treason than Germans. Even arrests for activities outside the Reich outnumbered German arrests between the middle of 1940 and the middle of 1941.

3.1.3 Resistance Resurgent (January 1941–December 1941)

Serious resistance to the regime revived in 1941. There was a steady growth of arrests, averaging around 39 per month, driven primarily by activities that took place outside the borders of pre-1938 Germany. Growth was most marked after the invasion of the Soviet Union with 80 arrests made in July 1941 alone. Austria was numerically the most important source of this resurgence of treasonous and high treasonous activities. Arrests for activities outside the Reich, however, lose their relative importance, declining to fairly low levels by the end of 1941 and remaining there for the rest of the war.

3.1.4 *Peak Resistance (January 1942–September 1943)*

In 1942, resistance continued to grow apace, reaching a monthly average arrest rate of 56. There were particularly significant spurts of arrests, especially in Austria, during and after the first Soviet winter counter offensive. In February 1942, 121 arrests were made, up to that point of time, as Fig. 3.5 shows, the most active month yet encountered and the second most active month the regime was to experience across its entire duration. Another significant, though less dramatic, burst of arrests occurred after defeat at Stalingrad, again felt most strongly in Austria, but this time clearly noticeable in Germany, too. Between January and March 1943, there were on average 73 arrests per month. Austria retained its numerical dominance throughout this period of resistance.

3.1.5 *Resistance and Regime in Decline (October 1943–May 1945)*

From October 1943, the volume of serious resistance and arrests declined back to lower levels, averaging at approximately 28 arrests per month between October 1943 and June 1944. The share of German arrests rose, placing it back on a par with Austrian in terms of numerical importance. The slow downward drift in arrest for treason and high treason was dramatically disrupted by the events of 20 July 1944 and its aftermath. In July and August 1944, as many as 289 arrests were made. July 1944, with 176 arrests, was the most active month with respect to treason and high treason of the National Socialist era. The impact of the 20 July plot was almost entirely confined to Germany, though. It had no discernible effect in Austria, and there was only a small uplift in arrests for activities outside the Reich. By September 1944, the spurt of arrests in Germany was over, too. Thereafter, the decline resumed and the number of arrests fell dramatically, averaging at around 17 per month in the quarter year between September 1944 and January 1945. In the final months of the war, the Nazi legal system began to collapse as Allied bombing and the advance of enemy forces onto home territory threw the workings of the People's Court into disarray. On 3 February 1945, an American raid on Berlin seriously damaged the buildings of the People's Court, killing its President, Roland Freisler (Evans 2008, p. 644). In the growing chaos of the regime's final months, cases of serious resistance increasingly were no longer processed through 'normal' legal channels, and when they were, record keeping of such proceedings became partial or simply non-existent. Towards the end of this period, the records of the People's Court ceased to reflect the patterns of serious resistance.

3.2 Places

Thus far, our analysis of serious resistance has been undertaken at the national level. In the remainder of this chapter, we turn our attention to the regional distribution of resistance activity. Tables 3.1 and 3.2 provide an overview of the regional spread of serious resistance in Austria and Germany, respectively. Austria is decomposed into its nine pre-*Anschluß* provinces, whereas the regions of Germany reported in Table 3.2 correspond to the 16 electoral constituency associations (*Wahlkreisverbände*) of the November 1932 Reichstag Election, plus Saarland, which at the time was a League of Nations mandate administered by France. It should also be noted that the numbers attributed in each region in both tables refer to the location where the defendant's resistance activities took place, not where they were arrested. These were not always the same. A resister who had engaged in activities in, say, Graz, for example, may have been apprehended by the authorities sometime later, perhaps while on the run, elsewhere inside or, indeed outside, the Reich.

Tables 3.1 and 3.2 report four pieces of information for each of the regions listed: their share of all national incidences of resistance activity, the total number of incidences that took place there, the expected level of incidence that might have been expected to have taken place there based on the region's share of the nation's voting population and the level of incidence per 100,000 registered voters in the region, a figure which can be taken to indicate not only the regional incidence rate but also the intensity of resistance in that location.[12]

Together these data allow us to explore with nonparametric statistics some aspects of the relationship between the regional distribution of resistance and populations. Applying chi-squared tests of the observed and expected number of arrests in both tables, thus, enables us to reject with 99% confidence the null hypothesis that the observed level of incidence in both Germany and Austria did not differ significantly from that expected from the voting population of the regions.[13] In other words, the regional level of incidence in Germany and Austria did not simply reflect the number of people of resistance age in those regions. Something else was driving the spread. Examination of the rank correlation of regional incidence and population size, however, finds that the relationship between the voting population size of a region and the level of incidence in that region can be described using a monotoni-

[12] The voting population is preferred here to the actual population as it is a closer match to the pool of potential resisters. The total population of a region, after all, includes significant numbers of infants and the elderly who were not likely to be engaged in serious resistance.

[13] The chi-squared test of independence is a nonparametric method of determining the goodness of fit between a set of observed values and those expected theoretically; see Greenwood and Nikulin (1996). Further details of all of the tests undertaken in this chapter are available from the authors upon request.

Table 3.1 Incidence of treason and high treason in Austria, 1933–1945 (by province of activity) (Ueberreuter 1931, pp. 3, 9)

	Share of incidence (percentage)	Level of incidence	Expected level of incidence	Incidence per 100,000 registered voters
Burgenland	1.95	37	74	22.92
Carinthia	4.53	86	99	39.80
Lower Austria	8.17	155	420	16.99
Salzburg	6.27	119	67	82.11
Styria	22.83	433	273	73.02
Tyrol	4.01	76	89	39.19
Upper Austria	6.01	114	245	21.44
Vienna	44.39	842	591	65.64
Vorarlberg	1.85	35	39	40.96
Austria as a whole	100	1897	1897	46.03

Notes: The expected level of incidence is calculated by multiplying a province's share of registered voters at the November 1930 National Election by the total level of incidence in Austria over the entire period. The resulting figure provides an indication of how much incidence there should have been had such incidences been determined purely by the size of the politically active population of the province (as measured by its number of registered voters)

cally increasing function in Austria, though not in the case of Germany.[14] That is to say, regions with larger populations tended statistically to have higher levels of incidence in Austria, but not in Germany. Further rank correlation analysis, however, shows that this relationship in Austria (and Germany) does not extend to incidence rates in regions (incidence per 100,000 voters).

Relative population size, thus, does not appear to drive the relative intensity of resistance activity in a region. In fact, the estimated correlation coefficient is close to zero and negatively signed in both cases, suggesting that not only is there little connection between the two variables, but whatever relationship does exist operates in the opposite direction to what would have been anticipated, namely, regions with larger populations have lower incidence (intensity of resistance). Interestingly, the German case differs from the Austrian in one respect. There is a very strong increasing monotonic relationship in Germany between regional incidence numbers and regional incidence rates. This finding suggests that those regions of Germany where there were larger numbers of arrests were also the regions where the intensity of resistance (measured by incidence per 100,000 voters) was high.

All in all, statistical analysis of Table 3.1 suggests that in Austria population size played at most a very broad background role in determining the regional distribution of serious resistance; it did not drive it. There are clearly some hotspots of Austrian resistance, most notably in Vienna, Styria and Salzburg (with incidence

[14] Rank correlation analysis provides statistical means of measuring ordinal associations. The rank correlation coefficient it engenders, thus, gauges the extent of similarity between two rankings. It is often used to determine the significance of the relation between variables; see Kendall (1970).

Table 3.2 Incidence of treason and high treason in Germany, 1933–1945 (by region of activity) (Falter et al. 1986, p. 73)

	Share of incidence (percentage)	Level of incidence	Expected level of incidence	Incidence per 100,000 registered voters
Baden-Wuerttemberg	6.23	291	355	8.42
Bavaria North West	2.01	94	245	3.95
Bavaria South East	3.88	181	268	6.94
Brandenburg (Berlin)	27.52	1285	154	85.86
Brandenburg II	1.78	83	418	2.04
East Prussia	0.75	35	148	2.44
Hessen	3.94	184	283	6.70
Lower Saxony and Bremen	4.88	228	323	7.25
Pomerania-Mecklenburg	1.46	68	195	3.59
Rhineland North	9.08	424	286	15.24
Rhineland South	3.85	180	248	7.48
Saarland	2.29	107	N/A	N/A
Saxony	12.66	591	379	16.03
Saxony-Anhalt and Thuringia	4.97	232	385	6.19
Schleswig-Holstein and Hamburg	7.97	372	211	18.11
Silesia	2.61	122	313	4.01
Westphalia	4.13	193	352	5.63
Germany as a whole	100	4670	4563	10.28

Notes: The expected level of incidence is calculated by multiplying a region's share of registered voters at the November 1932 Reichstag Election by the total level of incidence in Germany over the entire period. The resulting figure provides an indication of how much incidence there should have been had such incidences been determined purely by the size of the politically active population of the region (as measured by its number of registered voters). Saarland has been excluded from these calculation because, as a League of Nation's mandate administrated by France until 1935, it did not participate in German elections prior to 1933

rates per 100,000 ranging from 65 to 82), though every region, including provinces of relatively low activity such as Upper and Lower Austria, experienced an intensity of activity greater than that found in most parts of Germany. Indeed, in all of Germany, it was only in Berlin (Brandenburg) where incidence rates matched levels found in Austria.[15]

[15]The regional distribution of incidence in Austria given in Table 3.1 is broadly consistent with those provided by Neugebauer (2013, p. 237) for the location of acts of high treason by Austrians heard by the People's Court. Both sets of figures identify and rank in the same order the same four most active provinces. In each case, these four provinces together accounted for over 80% of all incidence in Austria. Moreover, statistical analysis of the independence of two series (chi-squared tests) indicates that they are not significantly different from each other. The differences that do exist between the two series are thus in all probability due to differences in the sampling.

Table 3.2 similarly indicates that the scale of a region's voting population in itself is also a very bad predictor of arrests and resistance activity in Germany. It also suggests that there was an even larger and distinctive heartland of serious resistance in Germany—greater Berlin—and a few other lesser, but still important, loci of activity, most notably in Schleswig-Holstein and Hamburg, Saxony and Rhineland North. As in Austria, such hotspots of resistance were not tied to regional population size. There were also a number of German regions where the levels of serious resistance activity were very low, notwithstanding their population size. Brandenburg II is the best example: very low levels of incidence for treason and high treason, yet the largest region in terms of population. Silesia, Bavaria North West, East Prussia and Pomerania-Mecklenburg were other notable regions of low resistance activity. In Chap. 5, when we analyse the contributions of different political groups and parties, these regional patterns of resistance will be investigated further.

As noted earlier in the chapter, serious internal resistance was not confined within the borders of the Reich.[16] Germans and Austrians were also arrested in numbers for serious resistance activities undertaken outside the Reich. Table 3.3 summarises the prevalence of such activities across 15 European countries.[17] Over 90% of such arrests for foreign-based activities occurred in four regions to which the regime's domestic political opponents had fled in significant numbers in the pre-war years: the Benelux countries and France (together accounting for 43.6% of all incidence outside the Reich), Czechoslovakia (34.4%), the Soviet Union (9.2%) and Denmark (5.1%).

Finally, narrowing the focus even further, Table 3.4 records the 15 most common places of serious internal resistance to the Third Reich. Together these locations accounted for just under two-thirds of all incidences of high treason and treason between 1933 and 1945. Not surprisingly, perhaps, this list illustrates both the importance of large urban centres to resistance groups and the broad geographical spread of their activities. Berlin and Vienna clearly dominate as true capitals of resistance, with more than a quarter of all incidences of resistance between them, but large cities from all parts of *Altreich* (German boundaries before the end of 1937) and Austria appear on the list as well, as does Prague and two cities never to fall within the borders of the Greater German Reich: Paris and Amsterdam. Serious resistance to the regime by Germans and Austrians was found in all parts of the Third Reich, as well as in the lands it had annexed or occupied.

---

[16] See Ueberschär (2011). Section V deals with Germans living in exile.

[17] It should be noted that Table 3.3 reflects only the serious resistance activities of those outside the Reich who could be apprehended by German authorities. It does not capture resistance activities by Germans and Austrians in places like the United Kingdom, which were beyond the Gestapo's reach.

Table 3.3 Incidence of treason and high treason outside the Reich, 1933–1945 (by country of activity)

Nation	Level of incidence	Share of incidence (percentage)
Belgium	81	7.26
Bulgaria	5	0.45
Czechoslovakia	387	34.68
Denmark	57	5.11
France	215	19.27
Greece	1	0.09
Hungary	41	3.67
Italy	1	0.09
Luxembourg	8	0.72
Netherlands	182	16.31
Norway	13	1.16
Poland	16	1.43
Romania	6	0.54
USSR	103	9.23
Yugoslavia	10	0.90

Note: Since some activities took place in more than one country, the total above sums to more than 100 %

Table 3.4 Most common places of resistance for defendants charged with treason or high treason

Rank	Location	Number of defendants	Share of defendants (percentage)
1	Berlin	1285	16.73
2	Vienna	842	10.96
3	Leipzig	333	4.33
4	Hamburg	322	4.19
5	Dresden	319	4.15
6	Prague	317	4.13
7	Graz	209	2.72
8	Duesseldorf	193	2.51
9	Paris	187	2.43
10	Munich	164	2.13
11	Amsterdam	147	1.91
12	Salzburg	140	1.82
13	Frankfurt am Main	116	1.51
14	Essen	114	1.48
15	Saarbruecken	104	1.35

3.3 Summary

This chapter has provided an overview of the temporal and geographical spread of serious resistance to the Nazi state by its citizens. Perhaps its most telling finding is that resistance was both widespread and persistent. Indeed, even when the regime was at the zenith of its powers, resistance remained an ever-present reality it had to confront.

Based on a close analysis of arrest figures for high treason and treason, this chapter has offered a new periodisation of serious resistance over the course of the regime's life. This approach has shed new light on aspects of the history of resistance. It has shown, for example, that serious resistance, while greatly damaged by the relentless persecution of the state and its own failures of the early years, was not moribund between 1935 and 1942, as commonly assumed. Neither defeat at Stalingrad nor the success of D-Day represented the crucial dividing line in the history of resistance. For the Reich as a whole, the upswing in serious resistance in fact began well and truly before that catastrophe on the Volga. Indeed, the period from February 1942 and the shock wrought by the first Soviet winter offensive emerges as perhaps the single most important climacteric in the history of serious resistance. In terms of arrests, 1942 was the year of peak resistance to the regime. The invasion of the Soviet Union, likewise, was a significant event, bringing in its wake a distinct surge in resistance activities, but it is noteworthy that this surge had begun to peter out by October–November 1941. Both the onset of Operation Barbarossa and the defeat at Stalingrad are less important than some might expect.

The picture unveiled in this chapter has other implications. While confirming the quantitative significance of Stauffenberg's assassination attempt in July 1944, it has also suggested that the 20 July conspiracy should be viewed as only one, albeit the single most significant, event, in the history of the internal resistance to Nazism. It is important to recognise, though, that this was an event that occurred against a backdrop of declining levels of serious resistance. Notwithstanding the wave of arrests and trials in the aftermath of the 20 July assassination attempt, there were in fact significantly fewer arrests for treason and high treason in 1944 than there had been in both of the 2 preceding years (1942 and 1943). Moreover, as only a quantitative analysis of the regime's whole life can reveal, there were other periods when there were very high levels of arrests, too, such as July–December 1933, December 1934–November 1935, February–December 1942 and January–September 1943. While they lacked the high drama of the 20 July plot, these periods nonetheless warrant the close attention of the historian of German and Austrian resistance.

The chapter has also provided evidence of multiple resistance experiences in the Third Reich. The story of resistance in Germany, Austria or beyond the Reich was the same in neither pattern nor scale. Furthermore, it should not be simply assumed that German-based activities always dominated quantitatively. For long stretches of time, such as for most of the period between 1940 and 1943, there were greater volumes of resistance in Austria than in Germany. Resistance was a highly regional, urbanised business, too. It was far from being a uniform, national phenomenon.

While Berlin and Vienna were overwhelmingly the most important centres of opposition to the regime, serious resistance occurred in all parts of the Reich. The patterns of incidence across regions suggest that the level and intensity of resistance experienced in any location were not simply reflections of population size. Yet, precisely what did drive the differences between regions remains unknown. Many important and fundamental questions about serious resistance, thus, remain open to researchers, none more so than in those relatively unchartered waters of causation: Why did resistance take firmer roots in some places than others? A research agenda suggested by the findings of this chapter would be to undertake a series of detailed comparative analyses of regions of high and low levels of resistance. Such an approach, especially when buttressed by empirics, offers, we believe, a promising avenue of research for historians of resistance, not to mention social scientists, wishing to explore and uncover the micro determinants and foundations of the level of resistance observed in any given location.

References

Primary Sources

Die Nationalratswahlen vom 9. November 1930. 1931. In *Statistisches Nachrichten 1931, Sonderheit.* Vienna: Ueberreuter.

Falter, Von Jürgen, Thomas Lindenberger and Siegfried Schumann. 1986. *Wahlen und Abstimmungen in der Weimarer Republik: Materialien zum Wahlverhalten 1919–1933.* Munich: Verlag C.H. Beck.

Topography of Terror Foundation ed. 2014. *The "House Prison" at Gestapo Headquarters in Berlin: Terror and Resistance, 1933–1945*, 3rd edition. Berlin: Topography and Terror Foundation.

Secondary Sources

Bailer, Brigitte and Gerhard Ungar. 2013. Die Zahl der Todesopfer politischer Verfolgung – Ergebnisse des Projekts. In *Opferschicksale: Widerstand und Verfolgung im Nationalsozialismus*, ed. *Dokumentationsarchiv des österreichischen Widerstandes*, 111–124. Freistadt: Plöchl Druck GmbH.

Benz, Wolfgang. 2014. *Der deutsche Widerstand gegen Hitler*. Munich: C.H. Beck.

Evans, Richard. 2008. *The Third Reich at War*. New York: Penguin.

Fest, Joachim. 1997. *Staatsstreich: Der lange Weg zum 20. Juli.* Munich: Taschenbuch Broschur.

Greenwood, Priscilla E. and Michael S. Nikulin. 1996. *A Guide to Chi-Squared Testing*. New York: Wiley.

Hanisch, Ernst. 1987. Gibt es einen spezifisch österreichischen Widerstand? In *Widerstand. Ein Problem zwischen Theorie und Geschichte*, Peter Steinbach, 163–176. Cologne: Wissenschaft und Politik.

Kendall, Maurice G. 1970. *Rank Correlation Methods*. London: Griffin.

Kershaw, Ian. 2015. *The Nazi Dictatorship: Problems and Perspectives of Interpretation*. London: Bloomsbury Revelations.

Lauf, Edmund. 1994. *Der Volksgerichtshof und sein Beobachter: Bedingungen und Funktionen der Gerichtberichterstattung im Nationalsozialismus*. Opladen: Springer Fachmedien Wiesbaden GmbH.

Marxen, Klaus. 1994. *Das Volk und sein Gerichtshof: eine Studie zum Nationalsozialistische Volksgerichtshof*. Frankfurt am Main: Klostermann.

McDonough, Frank. 2015. *The Gestapo: the Myth and Reality of Hitler's Secret Police*. London: Coronet.

Moorhouse, Roger. 2010. *Berlin at War*. New York: Basic Books.

Neugebauer, Wolfgang. 2013. Der österreichische Widerstand 1938–1945. In *Opferschicksale: Widerstand und Verfolgung im Nationalsozialismus*, ed. *Dokumentationsarchiv des österreichischen Widerstandes*, 233–272. Freistadt: Plöchl Druck GmbH.

Schlüter, Holger. 1995. *Die Urteilspraxis des nationalsozialistischen Volksgerichtshof*. Berlin: Duncker & Humblot.

Ueberschär, Gerd ed. 2011. *Handbuch zum Widerstand gegen Nationalsozialismus und Faschismus in Europa 1933–39 bis 1945*. New York: De Gruyter.

Wagner, Walter. 1974. *Der Volksgerichtshof im nationalsozialistischen Staat*. Stuttgart: Deutsche Verlags-Anstalt.

Chapter 4
Faces and Contexts

Serious resistance is, by its very nature, an acutely political act. After all, its intent is – or is at least deemed by the resisted regime to be – nothing less than a change to the existing political system. Understandably, actors in such struggles of resistance tend to be perceived primarily through the visor of their party or group's political ambition and aspirations; they are first and foremost seen as mere agents of the change they seek to usher in. The vast majority, of course, were deeply political and did passionately subscribe to the advancement of a particular ideology, theology or cause. In the next chapter, we will look at that political dimension of their resistance. That said, even the most committed of political activists was more than just a canvass for their belief systems. They were individuals, each with their own relationships, families, jobs, careers and a rich array of personalised experiences that also shaped who they were. Yet, with a number of notable exceptions, these aspects of their lives have largely been left unexplored in the literature.[1] It is as if, beyond their political actions and affiliations, the resisters themselves are unknown.

This chapter begins to redress this situation by attempting to put back some of the human face and context into the story of those who resisted the Nazi regime. Who were these resisters and what, if anything, was distinctive about them other than their willingness to resist and the political parties and organisations they joined? To address these questions, this chapter surveys the serious resister's experiences, identifies some of their key personal characteristics and considers, in addition to their religious and political affiliations, the significance of such features as their age, gender, family circumstances, ethnicity, occupation, education and criminal record.

---

[1] See, for example, a recent biography of Stauffenberg by Hoffman (2008). The 20 July plot has also been immortalised in several films, the latest of which was 'Valkyrie' in 2008. The deeds and key members of the White Rose are well known in both book and film; see Hanser (2012), 'Sophie Scholl: The Final Days' (2005) and Sturms (2013).

W. Geerling, G. Magee, *Quantifying Resistance*, Studies in Economic History,
DOI 10.1007/978-981-10-6008-3_4

4.1 Age

Figure 4.1 presents the age distribution of individuals indicted for treason and high treason between 1933 and 1945. The youngest at the time of indictment was just 16 years of age; the oldest is 77. This distribution is compared in Fig. 4.1 to that of the population of the Reich as a whole in 1939, as well as to that section of the German population, from whom age-wise resistance was most likely to have emerged (one might call this the pool of potential resisters). This latter distribution, made up of all inhabitants aged 16 or more, has also been presented because the inclusion of pre-teenagers in 1939, who constituted around a quarter of the German population, in the Reich's distribution skews the comparison between the two series significantly. The effect can be seen in the median age group of the different distributions, which is in the range 35 to <40 for those indicted and 30 to <35 for the Reich. However, if those younger than 16 are removed from the Reich's distribution, the median age of the rest of the Reich lies in the same band as those indicted (i.e. 35 to <40). Overall, though, even after allowance for the size of the pool of potential resisters is made, the distribution of age across the indicted differs. Serious resistance was overwhelmingly the game of young and middle-aged adults between the ages of 20 and 50, namely, individuals in the prime of their political lives for whom successful regime change, one would expect, could have brought significant long-term benefits. Eighty-five percent of all those tried for treason and high treason fell into this age bracket, compared to just 45% of the whole population (and 60%

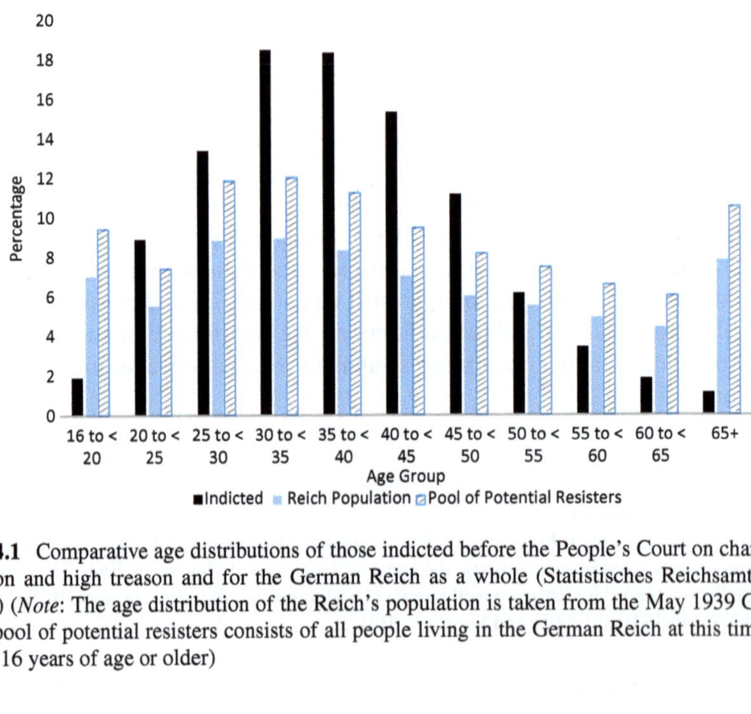

Fig. 4.1 Comparative age distributions of those indicted before the People's Court on charges of treason and high treason and for the German Reich as a whole (Statistisches Reichsamt 1940, p. 24) (*Note*: The age distribution of the Reich's population is taken from the May 1939 Census. The pool of potential resisters consists of all people living in the German Reich at this time who were 16 years of age or older)

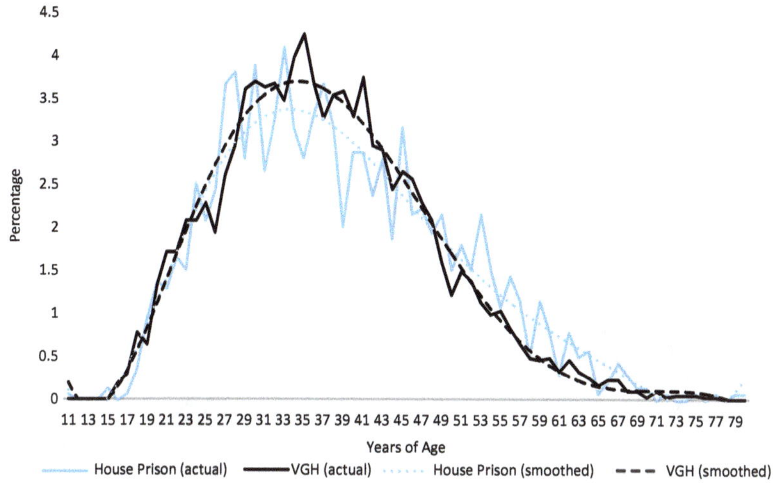

Fig. 4.2 Comparative age distributions of those indicted before the People's Court on charges of treason and high treason and those interrogated at the House Prison, 1933–1945 (Topography of Terror Foundation 2014, pp. 214–237) (*Note*: VGH stands for *Volksgerichtshof* (People's Court))

of the pool of potential resisters). By contrast, the very young and the over 50s were significantly under-represented (increasingly so with age for the elderly) among serious resisters. These findings may suggest that declining energies and capabilities, the fear of losing whatever position and status they had already attained in life and the unlikelihood that they would actually see personal benefit from political change themselves inclined many older Germans and Austrians to risk averseness.

Figure 4.2 provides another comparison. It plots the age distribution of those charged with treason and high treason against those interrogated in the House Prison, the main holding cells for the Berlin *Gestapo*. The House Prison data provide the most comparable coverage of ages to that given by the People's Court data. The comparison reveals a number of things. First, in both cases, there is a fairly normal distribution across the different ages with a slightly elongated tail on the older (i.e. right-hand) side of the distribution. Most interrogated by the *Gestapo* fell within the band between 20 and 50 years of age. As an eyeballing of both series in Fig. 4.2 readily indicates, there is strong similarity between the age distributions. In fact, the correlation coefficient between the two is of the magnitude of 0.94, suggesting very strong parallels. They are, of course, not identical. The main differences are (i) the much greater variability in the House Prison series and (ii) its larger share of older adults (40+ years old). This latter difference gives rise to the longer and thicker right-hand-side tail of the House Prison series. The presence of a somewhat larger proportion of older adults interrogated at the House Prison stems from the fact that the 20 July plotters quantitatively dominated the 1944 and 1945 observations for the series (unlike, as we saw in the last chapter, the People's Court series, where the 20 July conspirators represented a minority of indictments in those years). The key conclusion to be taken away from these comparisons is that, in terms of age

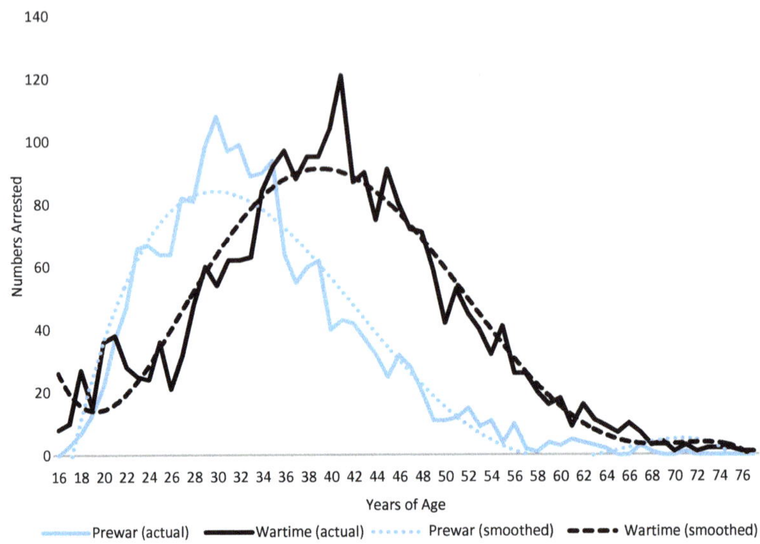

Fig. 4.3 Pre-war and wartime age distributions of those indicted before the People's Court on charges of treason and high treason

distribution, the People's Court indictment series is not only fairly normal but, in fact, more representative than that given by the House Prison series.

The age distribution of those arraigned before the People's Court, of course, did not remain

static across time. As Fig. 4.3 neatly illustrates, while the overall distribution retained its normality throughout, the curve itself shifted significantly to the right during wartime. In other words, the mean age of those indicted for treason and high treason was lower in peacetime than during the war (34 years of age compared to 37).

Table 4.1 amplifies the point. The mean ages of those indicted increased steadily across the course of the war, as did the standard deviation of age or, in other words, the extent of variation in the ages observed.[2] This pattern was most marked for males. Their average age rose from about 34 years of age in the pre-war period to as high as 45 by the final stage of the war. As Table 4.1 shows, there was an increase of more than 3 years in the average age of resisters across each of our periods. Concurrently, the standard deviation of age rose from 9 years in the pre-war period to 11 after D-Day.[3] The average age of females indicted likewise increased but not

[2] The median age of those indicted displayed the same pattern, too, growing from 32 years in the pre-war period to 37 in the early stages of the war, 39 between the invasion of Russia and Stalingrad, 42 between Stalingrad and D-Day and, finally, 43 in the final year of the war.

[3] When a data distribution is approximately normal, about 68% of its values are located within one standard deviation of the mean. The standard deviation values reported, therefore, mean that in the pre-war period, 68% of observed male resisters were aged between 25 and 43 years of age; yet by the end of the war, that range had shifted to 34 and 56 years of age. These figures indicate both an aging and greater diversification in the age profile of male resisters over time.

Table 4.1 Mean age of those indicted before the People's Court on charges of treason and high treason and those interrogated at the House Prison, by period and gender (standard deviation in parenthesis) (Topography of Terror Foundation 2014, pp. 214–237)

	People's Court indicted							House Prison Interrogated		
	Pre-war	Early war	Barbarossa to Stalingrad	Stalingrad to D-Day	Post D-Day	Wartime as a whole (1939–1945)	Overall (1933–1945)	Pre-war	Wartime	Overall (1933-45)
Male	34	37	39	42	45	40	38	37	43	38
	(9)	(9)	(10)	(11)	(11)	(11)	(10)	(11)	(12)	(12)
Female	32	39	37	37	39	38	36	40	43	40
	(9)	(11)	(13)	(12)	(10)	(12)	(11)	(12)	(8)	(12)
Total	34	37	39	42	44	40	37	37	43	39
	(9)	(10)	(10)	(11)	(11)	(11)	(10)	(11)	(12)	(12)

as precipitously (from 32 years in peacetime to 39 at the end of the war). Interestingly, the mean age of women tried for high treason and treason grew markedly by 7 years (to 39) in the early stages of the war, declined to 37 following Operation Barbarossa and then rose again to 39 after D-Day.[4] Variation in the spread of female ages also did not follow the same path as male, rising to relatively high levels (of around 13 years) in the 2 years following the invasion of Russia, then steadily dropping off after defeat at Stalingrad to just 10 years in the final months of the war. These statistics suggest that the nature and experience of female resisters differed significantly from male. For one thing, they suggest that war in Russia and its consequences drew women of a wide variety of ages into the serious resistance movement for the first time. Many of these new female resisters were younger than average.[5] Later in the chapter, we will return to the distinctive nature of female resistance.

Looking at the spread of ages by stage of life, Table 4.2 reveals that the share of both young and mature adults (18–40 year olds) among those indicted fell away starkly for both genders as the war progressed. Serious resistance steadily came to be dominated by mature and older adults. While it is true that juvenile resistance also become more common during wartime, as a phenomenon it still accounted for only a relatively small share of those at the time of their indictment (around 1.6%). Taken together, the increasing age of resisters reflected the fact that 18–40-year-olds

[4] The average female ages reported here follow a similar pattern to those for Austrians charged for high treason. Until 1941, this stood at 30 but increased sharply between 1941 and 1944 to around 40 years of age. The average age of Austrians appearing before the Higher Regional Court in Vienna on charges of high treason was 38: the same age for women overall for the Reich between 1939 and 1945; see Neugebauer (2013, pp. 236–7).

[5] This finding is implied from the change in the standard deviation of the distribution between the early war years and in the immediate aftermath of Operation Barbarossa (and before Stalingrad). In the early war years, the range of ages for women within one standard deviation of the mean was 28–50; between Barbarossa and Stalingrad, that range broadened at the bottom end to between 24 and 50 years of age.

Table 4.2 People's Court indictments and House Prison interrogations, by stage of life (in percentages) (Topography of Terror Foundation 2014, pp. 214–237)

	People's Court indicted				House Prison interrogated			
	Juvenile	Young adult	Mature adult	Older adult	Juvenile	Young adult	Mature adult	Older adult
Pre-war								
Male	0.2	24.0	55.6	20.2	0.3	20.2	46.0	33.5
Female	0.0	33.5	51.4	15.1	0.0	19.5	34.7	45.8
Total	0.2	24.9	55.2	19.7	0.3	20.1	44.7	34.9
Wartime								
Male	1.7	8.4	40.4	49.5	0.3	11.0	29.7	59.0
Female	1.4	21.5	38.8	38.3	0.0	0.0	50.0	50.0
Total	1.6	10.3	40.2	47.9	0.3	10.3	31.1	58.3
Overall (1933–1945)								
Male	0.4	16.0	47.1	36.5	0.3	18.1	42.2	39.4
Female	0.9	25.6	43.1	30.4	0.1	16.4	37.1	46.4
Total	0.5	17.2	46.6	35.7	0.3	17.9	41.7	40.1

Note: A juvenile is defined as someone up to and including 17 years of age; a young adult 18–27 years old; a mature adult 28–40 years old; and an older adult more than 40 years old

were over time ever more being drawn into military service or war work. For males, the shift in ages was also a product of the growing involvement of senior bureaucrats and military personnel (most of whom were older than 40) into serious resistance activities towards the end of the war. This aging profile of resistance also meant that the types of people being investigated by the *Gestapo*, and presumably the techniques employed for their detection and interrogation, must have correspondingly changed across the war. The typical male resister of the pre-war period was aged somewhere between 25 and 43; by late 1944, he was between 34 and 56 – on average a somewhat more mature individual.

In these regards, how does the People's Court data compare with the House Prison data for those interrogated by the *Gestapo* in Berlin? Looking closely at Table 4.2, a similar picture emerges, albeit with three noteworthy differences. First, the mean age of those interrogated was somewhat older than those indicted: 39–37 years of age. Second, while the mean ages of males were effectively the same (38 years) in both databases, the mean age of females interrogated was higher (40 compared to 36 years of age for those indicted). Third, the standard deviation in the age of those interrogated was greater (12 years compared to 10). Again, these differences appear to have emanated from the much longer and more pronounced right-hand tail (i.e. the greater preponderance of older adults) in the House Prison series. Table 4.2 neatly captures the relatively larger proportion of older adults, especially female (and correspondingly a smaller share of mature adults), who were interrogated at the House Prison.[6]

[6] Bailer and Ungar (2013, p. 116) also provide information on the age distribution of the Austrian victims of political persecution, which is likewise similar to that presented in Fig. 4.1. They found,

Table 4.3 presents a range of other features of those indicted, broken down by both stage of life and gender. Contrasting the figures there with information from the May 1939 census reveals some interesting aspects of the population of serious resisters (Statistisches Reichsamt 1940, pp. 24–6). It shows that a disproportionate share of resisters were Austrian. Austrians made up around 8.4% of the Reich's population but around a third of the total indicted for treason and high treason. Foreigners, the stateless and those of partial Jewish ancestry (representing 0.86, 0.16 and 0.15% of the Reich's population, respectively), while each relatively small in terms of their share of the total population of resisters, were also significantly more prone to resist than the general population.[7] The contribution of those with partial Jewish ancestry was especially notable among the young (16–27-year-olds). Nearly one in ten young adults arrested for treason or high treason had some Jewish ancestry: a rate of arrest that was more than 6000 times greater than their incidence in the population. Such statistics confirm that those ethnic groups persecuted or annexed by the Nazi regime, although often small numerically, did figure prominently in actions of serious resistance.

Bruno Baum, the son of a Jewish tailor and ironer born in 1910, was typical of this variety of young resister. Educated at a Jewish boy school in Berlin, he undertook an apprenticeship and then worked in the capital as an electrician from 1928 through to 1930. His political awareness, however, had developed earlier as a teenager. In 1926, he joined the Communist Youth Association of Germany (*Kommunistische Jugendverband Deutschlands* or *KJVD*) and Red Youth Front (*Rote Jungfront*), eventually becoming a full member of the Communist Party of Germany (*Kommunistische Partei Deutschlands* or *KPD*) in 1927 and Red Front Fighters' League (*Rote Frontkämpferbund*) in 1929. His activities in the Red Front saw him arrested and imprisoned by Weimar authorities. In 1928, he was admitted to the *KPD's* school for cadres and steadily became more engaged in political work. By 1933, he was the leader of the *KJVD* in Berlin, a role he was to leave at the end of 1934 in order to attend the International Lenin School in Moscow for a year. On returning to Germany, he once again resumed his underground work for the *KJVD*, for which he was finally rounded up and arrested in December 1935. Aged just 27

for example, that 79.3% of Austrian victims were aged between 26 and 55 years of age and 12.7% between the ages of 16 and 25. In our dataset, 81.6% and 17.1%, respectively, of those who appeared before the People's Court on charges of high treason and treason were in those two age ranges.

[7] A stateless individual was someone who could not claim the citizenship of any nation. One such stateless German resister was Ulrich Osche, the son of a Latvian father, who had successfully applied for German citizenship around 1923. When his father died, however, he, along with his mother and brother, who all held left-wing views, had their German citizenship stripped from them, in the process rendering them stateless. Continuing his leftist activities, between March 1933 and early 1934, he acted as an organizer for the Revolutionary Trade Union Opposition (*Revolutionäre Gewerkschafts-Opposition* or RGO), before deciding to flee to Holland in March 1934. Until finally apprehended in March 1936 at the age of 25, he worked for the *KPD* and *KJVD* mainly out of Amsterdam and Cologne, transporting illegal anti-regime pamphlets into Germany. He was sentenced to 15 years in a penitentiary; see Anklage 14 J 288/36, pp. 1–2; 11–19; Urteil 1H 48/36, pp. 18–25.

Table 4.3 Personal experiences and characteristics of the indicted, by stage of life and gender (in percentages)

Characteristic or experience		Juvenile	Young adult	Mature adult	Older adult	Male	Female	Overall
Education and training								
Highest educational attainment	Primary	66.7	74.4	84.4	85.4	83.0	80.8	82.9
	High school medium	4.8	4.9	2.8	1.9	2.9	2.8	2.9
	High school advanced	28.6	8.1	4.7	2.0	4.0	7.4	4.4
	Tertiary	0.0	12.5	8.1	10.8	9.9	9.0	9.8
Vocational training	Completed or in process	66.7	57.0	56.4	53.9	59.7	26.2	55.7
Citizenship and ethnicity								
Citizenship	German	33.3	70.2	65.5	57.1	63.7	59.3	63.1
	Austrian	66.7	24.4	30.1	38.7	31.7	37.0	32.4
	Foreign	0.0	4.8	2.5	1.6	2.6	2.2	2.6
	Stateless	0.0	0.7	1.9	2.6	2.0	1.5	1.9
Ethnicity	Partially Jewish	4.8	9.2	2.2	1.9	2.7	7.4	3.3
Experiences								
Personal and familial	Chronic health problem	0.0	6.5	8.9	11.8	10.1	4.6	9.5
	Spent time outside the Reich (in a private capacity)	4.8	12.4	13.7	14.4	13.6	14.0	13.7
	Engaged or married (*de jure* and de facto)	0.0	19.0	65.9	81.2	64.5	52.1	63.0
	Widowed	0.0	0.4	1.3	4.1	1.5	7.0	2.1
	Divorced	0.0	1.3	5.3	5.9	4.3	8.5	4.8
	Has children	0.0	6.9	24.8	40.4	27.3	26.7	27.2
	Came of age during 1918–1919 Revolution	0.0	0.0	28.4	28.3	24.3	16.8	23.4
	Came of age during the hyperinflation	0.0	4.3	58.4	13.4	33.7	26.0	32.7
	Both parents alive	71.4	NR	NR	NR	NR	NR	NR
	Immediate family engaged in anti-regime activities	28.6	16.1	14.0	15.8	9.9	51.4	15.1
	Served in WWI	0.0	0.0	9.6	53.6	27.0	0.0	23.6
	Served in WWII	0.0	5.9	4.1	4.2	5.0	0.0	4.4
	Family member killed in war	19.0	2.3	1.2	3.5	2.2	3.3	2.3

(continued)

Table 4.3 (continued)

Characteristic or experience		Juvenile	Young adult	Mature adult	Older adult	Male	Female	Overall
Religious	Motivated by strong religious convictions	28.6	10.0	5.9	8.4	7.7	6.4	7.6
	Roman Catholic	19.0	9.2	4.9	7.3	6.6	6.3	6.6
	Evangelical	0.0	0.5	0.9	1.0	1.0	0.2	0.9
	Mormon	9.5	0.3	0.0	0.0	0.1	0.0	0.1
Legal	Previous convictions	9.5	23.0	33.6	30.0	33.1	11.4	30.4
	Previous terms of imprisonment	4.8	20.0	28.9	26.0	28.6	9.4	26.2
	Previous major political convictions	0.0	4.0	8.1	9.5	8.5	3.5	7.9
	Previous minor political convictions:							
	Defamation	0.0	2.1	3.9	4.9	4.3	1.1	3.9
	Public disorder	0.0	2.7	4.4	2.0	3.5	0.9	3.2
	Resisting state authorities	0.0	1.7	2.3	1.7	2.1	0.6	1.9
	Illegal meetings and/or writings	0.0	6.3	6.6	4.9	6.3	3.1	5.9
	Malicious gossip	0.0	0.1	0.1	1.0	0.5	0.0	0.5
	Previous non-political convictions:							
	Property crimes	4.8	7.1	9.0	8.3	9.3	2.4	8.4
	Sexual offences	0.0	0.5	1.4	1.2	1.2	0.7	1.2
	Crimes of violence	0.0	5.1	6.2	5.2	6.4	0.4	5.6
	Other crimes	4.8	7.9	10.8	8.6	10.4	2.8	9.5
Organisational memberships (current or past)	Denied entry/expelled from Hitler Youth/League of German Girls	14.3	4.3	0.1	0.0	0.9	0.9	0.9
	Youth street gang	4.8	0.9	0.0	0.0	0.2	0.0	0.2
	Non-political and non-religious	9.5	20.5	16.6	11.1	15.4	14.4	15.3
	Trade union	0.0	12.9	31.9	40.9	34.3	13.4	31.7
	Free Corps/Association of Front Soldiers	0.0	1.5	3.7	6.3	4.8	0.4	4.2
	SA/SS	0.0	3.1	2.5	1.4	2.5	0.0	2.2
	Hiking group	9.5	4.8	2.0	0.4	2.0	1.3	1.9
	Austrian Home Guard	0.0	0.7	1.6	1.4	1.5	0.2	1.3
	Protecting the Austrian Republic	0.0	0.7	7.2	9.1	7.6	0.7	6.7

(continued)

Table 4.3 (continued)

Characteristic or experience		Juvenile	Young adult	Mature adult	Older adult	Male	Female	Overall
German political memberships (current or past)	Communist	4.8	39.1	42.6	30.7	39.4	24.9	37.6
	Social Democrat	4.8	13.6	12.2	18.0	15.2	9.6	14.5
	Socialist	0.0	5.6	6.3	8.7	7.2	5.3	7.0
	National-conservative	0.0	1.1	2.2	3.0	2.6	0.4	2.3
	Centre Party	4.8	3.6	1.2	1.6	2.0	0.2	1.8
	Protecting the Weimar Republic	0.0	2.1	2.8	3.1	3.2	0.0	2.8
	Anarchist	0.0	0.8	0.7	0.7	0.8	0.6	0.7
	Black Front	0.0	1.9	2.0	1.5	1.9	1.3	1.8
	Trotskyist	0.0	0.9	0.9	0.3	0.7	0.9	0.7
	German Nazi	0.0	0.9	2.6	3.3	2.8	1.1	2.6
Austrian political memberships (current or past)	Christian Social	4.8	0.3	0.5	1.5	0.8	1.5	0.9
	Communist	0.0	10.4	11.7	11.6	11.5	10.1	11.4
	Social Democrat	0.0	7.5	17.3	24.7	19.2	11.0	18.2
	Austrian Nazi	0.0	1.3	1.4	1.7	1.5	1.8	1.5
	Fatherland Front	19.0	4.5	9.0	15.0	10.9	7.2	10.4
	Legitimists	19.0	2.0	1.9	3.4	2.4	3.3	2.5
	Austrian national conservatives	4.8	1.5	1.1	1.7	1.4	1.3	1.4
	Austrian Pan-German	0.0	0.1	0.3	0.5	0.3	0.6	0.4

Note: An NR in the column means this information was not recorded by the authorities consistently

at the time, he was to be later sentenced by the People's Court to 13 years in a penitentiary for membership of an illegal organisation and participation in the production and importation of illegal writings and other subversive materials into Germany (Anklage 17 J 28/36, pp. 1–7; Urteil 2H 24/37, pp. 1–4).

Indicative of the disproportionately large number of foreigners who became involved in the German resistance was Meir Broser, the son of a Russian father and Jewish mother born in Tula, Russia, in 1891. At the age of six, he and his family moved to Berlin from Bessarabia. After finishing school, he began studying philosophy at a university in Berlin but, after one and a half semesters, had to relocate to Leipzig. He continued studying there until interned as an enemy alien at the beginning of the First World War. Released from detention in 1916, he found work as a warehouseman. In 1918, he was eventually able to return to university but, once again, had to discontinue, this time as a result of a lack of money. He moved to Stuttgart, where he gave Russian lessons for a while, before moving onto Lithuania. There, he became a teacher, eventually rising to become principal of the high school he was at. However, at the behest of his wife, a German he had married in 1918, they returned with their two children to Germany in 1925 with the intention of finishing his doctoral studies in Breslau. From 1927, he worked in Berlin as a language teacher. Although he had no history of political affiliations, he was arrested in September 1943 after having spent a number of weeks in a guest room with others who had. It was alleged by the authorities that he has been giving his associates Russian lessons, reading subversive materials, listening to the British Broadcasting Corporation (BBC), and providing other general assistance to the banned socialist organisation, the European Union (*Europäische Union*). For such activities, he was jailed for 2 years in a state prison (Anklage 11 J 36/44, pp. 1–4; Urteil 1H 64/44, pp. 1–2; 5).

Table 4.3 also reveals that among the serious resisters, one was more likely to find divorcees. They comprised 4.8% of the resister population but only 1% of the national population (1.4%, if we restrict the population to those 16 years and above who were of marriageable age). By contrast, widowers, who made up as much as 8.5% of the national population, were noticeably under-represented among the indicted. Compared to all Germans and Austrians of marriageable age, however, the rate of marriage among those charged with treason and high treason was very similar: 61.5% nationally to 63% among resisters.

Many of the other characteristics and experiences of the indicted reported in Table 4.3 follow expected patterns. Chronic health problems, travelling outside the Reich in a private capacity, the likelihood of being married and having had children all steadily grew with age. The share of resisters with previous convictions, both of the political and non-political variety, and spells of imprisonment also increased up until mature adulthood, after which they all diminished slightly, a diminution no doubt in part due to the broader age range captured by the older adult stage of life (viz. anyone 40 years or older). Attributes such as having had an immediate family member either engaged in anti-regime activities or killed in war and having served in World War II are shared fairly evenly across all age groups, with the exception of juveniles.

Each stage of life, of course, had its distinct features. The young adult category was the most German in its composition, although, interestingly, it also included among its numbers the largest share of foreigners and people of partial Jewish ancestry. It was also the best educated cohort, with the largest shares of tertiary- or high school-educated resisters. Not surprising for young adults, perhaps, it was also the stage of life most heavily engaged in hiking groups and other non-political and non-religious community-based organisations, such as sporting clubs and the Red Cross. Politically speaking, young adults were disproportionately communist, and less likely to be members of established social democratic or national-conservative political parties, or engaged in formal trade union activity. They were, however, more inclined to be driven by religious convictions, particularly Roman Catholicism, a fact also reflected in their strong representation in the Centre Party (*Zentrumspartei*).

In marked contrast to younger adults, mature adults, the largest age group of all, set the norm in most characteristics and attributes for the resister population. In the main, they were ethnic Germans, who were married and primary school educated. Probably, their most distinguishing characteristic was their heavy involvement in communist organisations, an engagement that had undoubtedly begun in their young adulthood. They were also ostensibly the least religiously motivated and the most likely to have had prior convictions and terms of imprisonment of all the age groups.

As for older adults, other than having the largest number of Austrians among them, they were fairly typical of the resistance population as a whole in terms of ethnicity or citizenship. Their educational levels were likewise comparatively normal, although they were the age group with the largest share of people with only a primary school education. Combined with their above-average share of tertiary-trained people, this finding strongly indicates the existence of two distinct groups of older resisters: the highly and the lowly educated. Such a dichotomy exists across all stages of life, of course, with the exception of juveniles, though it appeared most markedly in those older than 40. In terms of politics, older resisters tended to be less communist and more closely associated with social democratic, nationalist and conservative organisations than other stages of life. They were also the most likely to have been in a Nazi organisation or a trade union.

A novel feature of the analysis presented in Table 4.3 is its inclusion of a separate juvenile category. In the period 1933–1945, at least 69 juveniles at the time of their arrest were charged with high treason and arraigned before the People's Court. More often than not, in studies of resistance, such teenagers are either ignored or lumped together with young adults. It is not hard to understand why. Juveniles did not possess the means, know-how or experience to effect conspiratorial resistance in the mould of Stauffenberg, prominent religious figures or the White Rose. Therefore, when historians began to document the 'other Germany' after the war, juveniles who did not fit the idealised picture of resistance fighters were subject to criminalisation or ridiculed as naive or idealistic and their actions described as insignificant. More than 70 years after the conclusion of the war, there is still scant analysis of juvenile resistance in the Third Reich (Geerling et al. 2013, pp. 209–234).

Table 4.3, combined with work done elsewhere by the authors, gives a rare overview of the profiles of juvenile resisters charged with treason and high treason in Nazi Germany. While each individual was unique and there was notable variety in backgrounds and experiences, several characteristics and features appear in a significant number of cases. One can thus reasonably talk of a 'typical' juvenile resister: a single, healthy individual of German ethnicity, 17 years old, from a two-parent family. Among juvenile resisters, there was a higher than usual probability that other members of their family had either been engaged in anti-regime activities, killed in the war or, indeed, both. If not still a student, they were employed as a blue-collar worker, often as an apprentice. Their highest educational achievement was to have completed at least some high or trade school. Before being arrested for high treason, they tended not to have had prior charges laid against them nor had they ever been outside the Reich. Earlier in their life, they had often experienced some trouble with the authorities, either in entering or staying in the Hitler Youth or the League of German Girls, and if they subscribed to any formal organisations at all, they were a member of either a left-wing or Austrian nationalist organisation or held strong religious convictions, usually of a Roman Catholic or Mormon persuasion. These beliefs appear, in part, to have motivated their actions. They were opposed to the war and, if Austrian, as indeed two-thirds of them were, the *Anschluß*. They were not a lone resister but instead tended to resist in groups with other juveniles largely free of significant adult supervision. In short, the typical juvenile resister was a fairly 'normal' teenager in all but one respect: their opposition to National Socialism.

A good example of such a 'typical' juvenile resister was Friedrich Lachnit, an apprentice boat builder from Vienna. He joined a Social Democratic youth organisation (*Kinderfreunde*) as a child and remained a member until the *Anschluß* in 1938. He then joined the Hitler Youth but left disillusioned a year later. During the war, he joined a Communist Youth Association of Austria (*Kommunistische Jugendverband Österreichs* or *KJVÖ*) cell in Vienna. In the period from 1941 to 1942, his cell sent anti-war newspapers and leaflets to *Wehrmacht* personnel on the Eastern Front. Six members of this cell were arrested (three of whom were juveniles) and charged with high treason, including belonging to an illegal organisation, the circulation of illegal writings, separatism, undermining the fighting spirit of the German armed forces and treason. Lachnit was sentenced to death by the People's Court and later executed. The two other juvenile members of the group received penitentiary sentences of 10 years and 5 years, respectively (BA, R 3017/alt R 60 II/269, pp. 1–2).

While it is accurate to assert that, in comparison to those whose goal was to assassinate Hitler or stage *coup d'états*, juvenile resistance tended to be marginal, spontaneous and lacking in clear political direction and strategy, close analysis of the records of those charged with treason and high treason demonstrates that their actions were far from being random, pointless or simply the by-product of common juvenile criminality. Rather, theirs was a distinct form of resistance, worthy of separate recognition, respect and academic study.

4.2 Gender

Another distinguishing feature within the population of serious resisters was gender. 12.3% of those arrested and indicted for treason and high treason were female, a share considerably lower than the female share of the Reich's population, which in May 1939 stood at 51.2%. It is a figure, however, that is broadly in line with the findings of regional studies of resistance. For example, according to Christl Wickert (1997, p. 112), who studied a random sample of 2000 *Gestapo* files and 400 court files from the Special Courts, Higher Regional Courts and People's Courts in Duesseldorf and Essen, 'approximately 15 per cent of known resistance fighters were women'. Similarly, Neugebauer (2013, pp. 236–7) found that 12.27% of Austrians charged with high treason before the Higher Regional Courts of Vienna and Graz and 11.3% before the People's Court were women. Bailer and Ungar (2013, p. 119) provided a somewhat lower percentage for the admittedly much broader category of female victims of political persecution in Austria: just 7%. They did note, however, that 19% of individuals appearing in daily reports from the *Gestapo* headquarters in Vienna were female, a figure that can be contrasted with the percentage of women interrogated by the *Gestapo* at the House Prison in Berlin, which stood at 10.1% (Topography of Terror Foundation 2014, pp. 214–237). A variety of sources, thus, confirm that female participation rates in serious resistance activities were far lower than male.

Until fairly recently gender-specific considerations had been largely excluded from studies of resistance or confined to a peripheral analysis of its place in the private sphere. Empirical and detailed archival work has begun altering the picture, however.[8] Wickert (1994c, pp. 13–14), for example, has importantly shown how women's defence of their living spaces was both a central motivation for their involvement in resistance and a major reason for the bias in their activities towards everyday acts of resistance and ideological dissent.

Traditional female roles and stereotypes, moreover, further restricted female engagement in the public sphere. As a consequence, Neugebauer (2013, p. 238) has argued that female resisters were typically considered by the Nazi authorities as less dangerous than their male counterparts. Such a presumption proved a mixed blessing. Women were certainly under-represented in active political resistance but were often more able than men to make up for it by playing an active role in the performance of more supportive tasks of resistance, such as providing food, shelter and moral comfort for those on the run; smuggling documents, weapons and money; acting as go-betweens; and setting up contacts and meetings. Charlotte Garske, a primary school-educated clerk from Berlin, for example, ran with her second husband a very successful safe house for *KPD* functionaries between 1934 and February 1943. Just 28 years old when she started the venture, she had only become politically active a short time before. Hitler's ascension to power had radicalised her. Among the guests to frequent her safe house were well-known figures of the communist resistance, Eugen Schwebinghaus (1934–1935), and the *KPD* Central Committee member, Wilhelm Knöchel (1942–1943). The People's Court sentenced her and her husband to death in 1943. Females, like Charlotte Garske, could also take advantage of their relative anonymity to circulate leaflets in and around their neighbourhoods and in shopping centres (Anklage 10 J 573/43, pp. 1–2; Urteil 2H 160/43, pp. 1–2).

Table 4.3 adds weight to these conclusions. It shows that, relative to males, far fewer women indicted for treason and high treason had had previous convictions for political activities. From the perspective of the authorities, they were often, at least initially, relatively invisible. Moreover, in many cases, female resistance was related more to personal everyday experiences, circumstances and relationships than politics. It was not uncommon for women to have been initially drawn into resistance in order to support others known to them. More than half had immediate family members engaged in anti-regime activities, while a disproportionately large share had

[8] The research inter alia of Frauke Geyken, Claudia Koonz, Ingrid Strobl, Jill Stephenson and Christl Wickert has been especially useful in shaping new perspectives; see, for example, Geyken (2014). For Koonz (1991, pp. 49–65). For Stephenson (2001, p. 45) and (2016, p. 20). For Strobl (1993, pp. 57–60). For Wickert (1994a, pp. 200–25), (1994b, pp. 411–26), (1994c) and (1997, pp. 101–113).

had a family member killed in the war. Female resisters were also not as heavily engaged as males in German political organisations or trade unions, with the possible exceptions of the *KPD* and Trotskyists parties, such as the International Communists of Germany (*Internationale Kommunisten Deutschlands* or *IKD*). In fact, communist groups provided one of the few vehicles available to women to participate in organised political resistance. But it came at a cost. According to Wickert (1997, p. 101), after Jewish and Gypsy women, female communists were among the most severely persecuted women in Nazi Germany. By contrast, in Austria, women appear to have been relatively more actively involved in political parties and not just in the Communist Party of Austria (*Kommunistische Partei Österreichs* or *KPÖ*) and Social Democratic Party of Austria (*Sozialdemokratische Partei Österreichs* or *SPÖ*) but also on the right of the political spectrum, where some played significant roles in the Christian Social Party (*Christlichsoziale Partei* or *CSP*), and the Legitimist, Austrian Nazi and pan-German movements.

Anna Stöger, for example, a casual shoe factory worker in Vienna, who had been born in the Lower Austrian market town of Perchtoldsdorf in 1906, was a committed member of both the *SPÖ* and an Austrian trade union between 1922 and 1934. In 1934, she joined the Fatherland Front and then, following her late, first husband — who had been a devoted communist — became active as the cashier in a *KPÖ*-organised group that provided arrested resisters and their families with financial support. Detained by the authorities in February 1943, she was sentenced later that year to a 15-year term in a penitentiary (Anklage 7 J 422/43, pp. 1–2; 6–7; Urteil 5H 122/43, pp. 1–3; 10–11).

As discussed earlier in the chapter, female resisters on average tended to be slightly younger than their male counterparts. They were also less likely to be married (almost half were single), though they were more prone to be either widowed or divorced. Ethnically, they were far more likely to have some Jewish ancestry. In terms of educational achievements, theirs were similar to male resisters, though it was more common for them to have progressed to the advanced levels of high school. Indeed, the share of women resisters with advanced high school or tertiary education was greater than among male resisters: 16.4–13.9%. Far fewer women pursued vocational training, however. Together, these figures suggest that, overall, the academic attainments of female defendants may on average have been slightly higher than their male co-resisters.

4.3 Occupation

Tables 4.4, 4.5 and 4.6 provide an overview, from a range of perspectives, of the occupational backgrounds of serious resisters. Above all else, they show the overwhelming dominance of the blue-collar worker, a branch of employment which accounted for more than half of those charged with treason and high treason. Service workers, full-time political activists, and white-collar workers were also prominently represented. Together, these occupational categories provided more than 85% of the serious resisters arrested and indicted. As Table 4.5 records, the most

Table 4.4 Occupational distribution of the indicted, by period (in percentages)

Occupation	Pre-war	Early war	Barbarossa to Stalingrad	Stalingrad to D-Day	Post D-Day	Wartime as a whole	Overall
Student	1.4	3.5	0.7	3.3	0.7	1.9	1.7
No particular occupation	3.4	5.0	4.8	7.7	11.0	6.6	5.2
Blue-collar worker (skilled and unskilled)	56.6	41.3	57.5	54.1	36.8	50.2	53.0
White-collar worker (excluding high-ranked state official)	7.8	9.1	8.0	5.3	8.6	7.6	7.7
Professional (excluding military)	3.7	9.1	2.9	6.0	9.6	6.0	5.0
Business or property owner	1.7	2.6	1.0	1.5	2.0	1.6	1.6
High-ranked state official	0.2	1.5	0.0	1.8	6.4	1.8	1.1
Engaged in full-time political work	8.4	12.7	8.6	3.8	6.9	7.8	8.0
Enlisted soldiers or non-commissioned military officer	0.3	0.2	0.8	1.4	1.2	0.9	0.6
Commissioned military officer	0.0	0.0	0.1	0.0	4.7	0.8	0.5
Service worker	16.6	14.9	15.5	15.2	12.3	14.8	15.6

common jobs held by resisters were a mixture of skilled and unskilled working-class vocations; most numerous among their ranks were mechanics, unskilled labourers, retail workers and (left-wing) political cadres. Each of these vocations accounted for in excess of 5% of all resisters indicted. By contrast, all of the business owners, professionals, high-ranking officials and members of the military (both senior and rank and file) combined comprised well less than 10% of resisters.

Table 4.4 also highlights the relative stability of this occupational distribution. The most discernible changes over time were the slightly greater involvement of both students and professionals in the early stages of the war and from the beginning of 1943; the increasing share of enlisted and non-commissioned military following the invasion of the Soviet Union, a process which reached its peak in the aftermath of the defeat at Stalingrad; and from mid-1944, the rapidly growing contribution of what may be called the 'establishment' (senior military officers, high-ranked state officials and business and property owners), which in this final period of the war attained significant levels for the first time. Yet, even in the tumult of the post D-Day environment, individuals from these predominantly middle- and upper-class occupations made up only 12.6% of those arrested and indicted for treason and high

Table 4.5 Most common occupations, by stage of life, gender and overall (with percentages in parenthesis)

	Juveniles	Young adults	Mature adults	Older adults	Male	Female	Overall
1	Apprentice (28.6)	Unskilled labourer (10.5)	Political cadre (10.2)	Mechanic (7.9)	Mechanic (9.7)	Clerk (9.8)	Mechanic (8.5)
2	Student (19.0)	Mechanic (9.9)	Unskilled labourer (8.7)	Political cadre (7.2)	Unskilled labourer (8.1)	Unskilled labourer (8.1)	Unskilled labourer (8.1)
3	Unskilled labourer (14.3)	Student (7.3)	Mechanic (8.5)	Unskilled labourer (6.0)	Political cadre (8.5)	Stenographer (6.4)	Political cadre (8.1)
4	Retail worker (14.3)	Retail worker (6.4)	Retail worker (5.4)	Retail worker (4.6)	Retail worker (5.2)	Retail worker (6.1)	Retail worker (5.3)
5	Fitter (9.5)	Clerk (5.5)	Construction worker (3.1)	Fitter (3.2)	Clerk (3.3)	Political cadre (4.8)	Clerk (4.1)
6	Mechanic (4.8)	Political cadre (4.4)	Carpenter (2.8)	Clerk (3.1)	Construction worker (3.1)	Student (3.1)	Construction worker (2.9)
7	Baker (4.8)	Construction worker (3.3)	Fitter (2.7)	Construction worker (2.4)	Fitter (3.0)	Tailor (2.2)	Fitter (2.7)
8	Chimney sweeper (4.8)	Typographer (3.2)	Clerk (2.2)	Carpenter (2.3)	Carpenter (2.1)	Teacher (2.2)	Carpenter (1.8)
9	–	Carpenter (2.7)	Miner (1.9)	Miner (1.6)	Typographer (2.0)	Cleaner (2.0)	Typographer (1.8)
10	–	Engineer (2.4)	Engineer (1.8)	Tailor (1.5)	Engineer (1.9)	Nurse (1.5)	Student (1.7)

Table 4.6 Occupational distribution of the indicted, by stage of life and gender (in percentages)

	Juvenile	Young adult	Mature adult	Older adult	Male	Female	Overall
Occupation							
Student	19.0	7.3	0.7	0.0	1.5	3.1	1.7
No particular occupation	0.0	2.7	4.5	7.5	0.8	36.6	5.2
Blue-collar (skilled and unskilled)	38.1	55.9	54.4	49.9	58.0	17.1	53.0
White-collar (excluding high-ranked state official)	4.8	11.5	7.3	6.5	6.0	19.3	7.7
Professionals (excluding the military)	0.0	4.7	4.6	5.8	5.0	4.8	5.0
Business or property owner	0.0	0.1	1.6	2.4	1.6	2.0	1.6
High-ranked state official	0.0	0.0	0.4	2.4	1.2	0.2	1.1
Engaged in full-time political work	0.0	4.4	10.2	7.2	8.5	4.8	8.1
Enlisted soldiers or non-commissioned military officer	0.0	0.9	0.5	0.6	0.7	0.0	0.6
Commissioned military officer	0.0	0.1	0.2	1.0	0.5	0.0	0.5
Service worker	38.1	12.4	15.5	16.8	16.1	12.0	15.6
Employment Status							
Employed	95.2	71.8	76.1	79.6	79.6	56.2	76.7
Unemployed	0.0	19.3	13.2	6.3	12.8	4.2	11.7
Not in the labour force	4.8	8.9	10.7	14.1	7.6	39.6	11.6

treason. The 20 July plot notwithstanding, the occupational structure of serious resistance in this period, as in all others, remained overwhelmingly working class.

Tables 4.5 and 4.6 examine the occupational spread from two other perspectives: stage of life and gender. They once again show a stable occupational structure across age groups (with the exception of juveniles) with the share of blue-collar and service workers remaining at around 50 and 15%, respectively. There were, of course, variations between stages of life. Juveniles were unsurprisingly more likely to be students and, when in the workforce, apprenticed or employed as unskilled labourers or retail workers. In young adulthood, many continued to study, although most had by then become blue-collar workers, especially of the unskilled variety, although some did find employment in more skilled and craft occupations, most commonly as mechanics, typographers, carpenters and engineers. Office-based clerical positions were also a common pastime. Nearly one in five young adults, however, was unemployed at time of arrest: the highest rate of unemployment of all age groups. Mature adults likewise experienced high rates of unemployment and were also the most engaged in full-time political work. Unskilled labourer, mechanic and retail worker continued as their most common occupations. Older adults were also typically blue-collar and service industry workers, although a sizeable minority (11.6%) held more substantial positions in society as either business owners, professionals, high-ranking bureaucrats or military officers. They were also more likely than any other age group not to be in the labour force or to be described as having no particular occupation.

Occupational data also sheds further light on the distinctiveness of female resistance. Female resisters were more inclined to be students or not to be in the labour force. While large numbers of women were low-skilled workers, they were on the whole less blue-collar than their male counterparts. Instead, they were to be found more frequently in white-collar roles such as clerical work or stenography or to be employed in traditional female occupations like nursing, teaching, domestic service, cleaning and tailoring (Mouton 2010, pp. 948–49).[9] And while there were significantly fewer female political cadres, state officials and military personnel – a reflection of their almost complete exclusion from the upper levels of the politics, government and defence of Nazi Germany – it is noteworthy that women resisters were equally likely as men to be a professional and, in fact, were more frequently described in official records as business and property owners. All in all, while still predominately working class, the occupational spread of female resisters appears to have been more socioeconomically diverse.[10]

4.4 Conclusions

This chapter has provided an overview of some of the key personal and environmental characteristics of those who participated in serious resistance to the Nazi regime. It has served to remind us that as a group they were more than just the politics to which they subscribed. Moreover, the chapter has described a range of interesting

[9] In this regard, Minna Klann was typical. After attending her primary school in a village in the Holstein region of Northern Germany, she worked as a maid until, at the age of 20, she married Erich Klann, a mechanic. They had three children together. Family responsibilities and chronic lung problems forced her from the workforce. She, however, remained politically interested, joining the *KPD* in 1926. At the time of her arrest in October 1935, she was highly active in the Luebeck branch of the party, collecting membership dues, distributing flyers and newspapers and attending meetings. For her role in maintaining what was by then an illegal organisation and distributing anti-regime material, she received an 8-year term of incarceration in a penitentiary. She was 36 years old; see Anklage 15 J 162/36, pp. 1–5; Urteil 1H 51/36, pp. 1–5.

[10] This occupational diversity chimes with research on female resistance and opposition undertaken by others. Wickert (1994a, p. 206), for example, has estimated that 19.1% of the women arrested in Duesseldorf for oppositional activities came from families of middle officialdom (*mittleren Beamtentum*) or had learned a commercial trade. 34.3% were of working-class origins. The corresponding figures for Essen were 10 and 25%, respectively.

phenomena: inter alia, the steady aging of resisters over the course of the regime, especially during the war; the dominance of blue-collar labour in resistance movements; and the small, though disproportionate, contribution of foreigners, the stateless and people of partial Jewish ancestry in the struggle to undermine Hitler's state from within.

Furthermore, we have seen that a very wide range of people from different occupations and backgrounds risked their lives by engaging in serious resistance. Opposition to the regime cut across all classes, spiritual beliefs and ethnicities. This fact notwithstanding, one of the strongest finding of this chapter has been the overwhelming numerical dominance of the working classes, both skilled and unskilled, in serious resistance activity. By contrast, high-ranking officials, military officers, business owners and professionals together represented only a relatively small share (8.2%) of those arrested for treason and high treason. In terms of overall resistance activity, theirs represented the exception more than the norm.

This chapter has also demonstrated the meaningfulness of analysing resistance by stage of life and gender. Experience of resistance could differ significantly by both age and sex. In particular, our analysis has highlighted the distinctiveness of two often overlooked types of resistance: juvenile and female. These two categories of resistance had much in common. Compared to the majority of resisters, who were adult males, women and teenagers were less political, less blue collar, less engaged in the formal labour force and seemingly more motivated to resist as a result of personal experiences and connections. Their lower participation rates and narrower range of activities reflected the fact that women and non-adults had fewer opportunities on the whole to engage in planned organisation-based resistance than adult males. As a result, their resistance often expressed itself more in non-violent acts within the private or semiprivate spheres. It is to be hoped that this chapter, by shedding more light on both, has helped to strengthen the foundations of a still emerging, albeit very different and more realistic, perspective on juvenile and female resistance, a perspective that sees both as the distinct resistance phenomena they were rather than the exceptional, random and essentially trivial acts they have far too often been assumed to be.

Official records reveal that as many as one in every 12 individuals charged with treason or high treason by the regime was actively involved in the day-to-day running and leadership of oppositional political parties. Given the intensely political nature of resistance, this is, of course, far from surprising. In the next chapter, we turn our attention to this political dimension and map the role played by different political groups and organisations in fermenting, initiating and executing serious resistance to the National Socialist state.

References

Primary Sources

(A) Archival Files

(i) Widerstand als Hochverrat

Anklage 14J 288/36 In: *Nationalsozialismus, Holocaust, Widerstand und Exil 1933-1945*. Online-Datenbank. De Gruyter. Accessed May 23, 2013. http://db.saur.de/DGO/basicFullCitationView.jsf?documentId=wh252

Anklage 15J 162/36 In: *Nationalsozialismus, Holocaust, Widerstand und Exil 1933-1945*. Online-Datenbank. De Gruyter. Accessed Apr 27, 2013. http://db.saur.de/DGO/basicFullCitationView.jsf?documentId=wh1026

Anklage 17J 28/36 In: *Nationalsozialismus, Holocaust, Widerstand und Exil 1933-1945*. Online-Datenbank. De Gruyter. Accessed May 24, 2013. http://db.saur.de/DGO/basicFullCitationView.jsf?documentId=wh331

Anklage 7J 422/43 In: *Nationalsozialismus, Holocaust, Widerstand und Exil 1933-1945*. Online-Datenbank. De Gruyter. Accessed Feb 12, 2014. http://db.saur.de/DGO/basicFullCitationView.jsf?documentId=wh2835

Anklage 10J 573/43 In: *Nationalsozialismus, Holocaust, Widerstand und Exil 1933-1945*. Online-Datenbank. De Gruyter. Accessed Oct 11, 2013. http://db.saur.de/DGO/basicFullCitationView.jsf?documentId=wh2563

Anklage 11J 36/44 In: *Nationalsozialismus, Holocaust, Widerstand und Exil 1933-1945*. Online-Datenbank. De Gruyter. Accessed Oct 7, 2013. http://db.saur.de/DGO/basicFullCitationView.jsf?documentId=wh719

Urteil 1H 48/36 In: *Nationalsozialismus, Holocaust, Widerstand und Exil 1933-1945*. Online-Datenbank. De Gruyter. Accessed May 23, 2013. http://db.saur.de/DGO/basicFullCitationView.jsf?documentId=wh253

Urteil 1H 51/36 In: Nationalsozialismus, Holocaust, Widerstand und Exil 1933-1945. Online-Datenbank. De Gruyter. Accessed Apr 27, 2013. http://db.saur.de/DGO/basicFullCitationView.jsf?documentId=wh1027

Urteil 2H 24/37 In: *Nationalsozialismus, Holocaust, Widerstand und Exil 1933-1945*. Online-Datenbank. De Gruyter. Accessed May 24, 2013. http://db.saur.de/DGO/basicFullCitationView.jsf?documentId=wh330

Urteil 2H 160/43 In: *Nationalsozialismus, Holocaust, Widerstand und Exil 1933-1945*. Online-Datenbank. De Gruyter. Accessed Oct 11, 2013. http://db.saur.de/DGO/basicFullCitationView.jsf?documentId=wh2564

Urteil 5H 122/43 In: *Nationalsozialismus, Holocaust, Widerstand und Exil 1933-1945*. Online-Datenbank. De Gruyter. Accessed Feb 12, 2014. http://db.saur.de/DGO/basicFullCitationView.jsf?documentId=wh3412

Urteil 1H 64/44 In: *Nationalsozialismus, Holocaust, Widerstand und Exil 1933-1945*. Online-Datenbank. De Gruyter. Accessed Oct 7, 2013. http://db.saur.de/DGO/basicFullCitationView.jsf?documentId=wh2757

(ii) Bundesarchiv

BA, R 3017/alt R 60 II/269, 1–2

(B) Other

Statistisches Jahrbuch für das Deutsche Reich 1939. 1940. Berlin: Statistisches Reichsamt.
Topography of Terror Foundation ed. 2014. *The "House Prison" at Gestapo Headquarters in Berlin: Terror and Resistance, 1933–1945*, 3rd edition. Berlin: Topography and Terror Foundation.

Secondary Sources

Bailer, Brigitte and Gerhard Ungar. 2013. Die Zahl der Todesopfer politischer Verfolgung – Ergebnisse des Projekts. In *Opferschicksale: Widerstand und Verfolgung im Nationalsozialismus*, ed. *Dokumentationsarchiv des österreichischen Widerstandes*, 111–124. Freistadt: Plöchl Druck GmbH.
Geerling, Wayne, Gary B. Magee and Robert Brooks. 2013. Faces of Opposition: Juvenile Resistance, High Treason and the People's Court in Nazi Germany. *Journal of Interdisciplinary History* 44: 209–234.
Geyken, Frauke. 2014. *Wir standen nichts abseits: Frauen im Widerstand gegen Hitler*. Bonn: C.H. Beck.
Hanser, Richard. 2012. *A Noble Treason: The Story of Sophie Scholl and the White Rose Revolt Against Hitler Vs the Revolt of the Munich Students Against Hitler*. San Francisco: Ignatius Press: San Francisco.
Hoffman, Peter. 2008. *Stauffenberg: A Family History*, 3rd edition. Quebec: McGill-Queen's University Press.
Koonz, Claudia. 1991. Choice and Courage. In *Contending with Hitler: Varieties of German Resistance in the Third Reich*, ed. David C. Large, 49–65. Cambridge: Cambridge University Press.
Mouton, Michelle. 2010. From Adventure and Advancement to Derailment and Demotion: Effects of Nazi Gender Policy on Women's Careers and Lives. *Journal of Social History* 43 (4): 945–971.
Neugebauer, Wolfgang. 2013. Der österreichische Widerstand 1938–1945. In *Opferschicksale: Widerstand und Verfolgung im Nationalsozialismus*, ed. *Dokumentationsarchiv des österreichischen Widerstandes*, 233–272. Freistadt: Plöchl Druck GmbH.
Stephenson, Jill. 2001. *Women in Nazi Germany*. New York: Routledge.
Stephenson, Jill. 2016. Women and Protest in Wartime Nazi Germany. In *Protests in Hitler's "National Community": Popular Unrest and the Nazi Response*, eds. Nathan Stoltzfus and Birgit Maier-Katkin, 18–37. New York: Berghahn.
Strobl, Ingrid. 1993. Sag nie, du gehst den letzten Weg. In *Widerstand der jungen Generation im NS-Staat: Dokumentation einer Tagung vom 17–19. Juli 1992 im Adam-von-Trott Haus*, 57–60. Berlin: Evangelisches Bildungswerk, Haus der Kirche.
Sturms, Frank. 2013. *Die Weiße Rose: Das Schicksal der Geschwister Scholl*. Frankfurt am Main: Marix Verlag.
Wickert, Christl. 1994a. Frauen im Hintergrund—das Beispiel von Kommunistinnen und Bibelforscherinnen. In *Das 'andere' Deutschland im Widerstand gegen den Nationalsozialismus: Beitrage zur politischen Überwindung der nationalsozialistischen Diktatur im Exil und im Dritten Reich*, 1st edition, eds. Helga Grebing and Christl Wickert, 200–25. Essen: Klartext.
Wickert, Christl. 1994b. Frauenwiderstand und Dissens im Kriegsalltag. In *Widerstand gegen den Nationalsozialismus*, eds. Peter Steinbach and Johannes Tuchel, 411–426. Berlin: Akademie Verlag GmbH.

Wickert, Christl. 1994c. *Frauen gegen die Diktatur: Widerstand und Verfolgung im nationalsozialistischen Deutschland.* Berlin: Hentrich.
Wickert, Christl. 1997. Women between Dissent and Resistance. In *Encyclopaedia of German Resistance to the Nazi Movement*, eds. Wolfgang Benz and Walter H. Pehle, 101–113. New York: Continuum.

Chapter 5
Groups and Organisations

For many of its supporters, one of the attractive features of National Socialism was undoubtedly the clarity and sense of purpose it appeared to offer. In times of great uncertainty, when all else seemed to have failed, a political philosophy that promised to restore certainty and confidence through strong leadership, discipline and the implementation of easily understood, populist solutions to pressing national questions was bound to appeal. But what was unknown to many at the time of their first embrace of National Socialism was that such promised certainty came with costs: war, genocide, intolerance and stultifying social conformity to name just some of the most important. Moreover, this was not an ideology that sought commonalities and unity; rather it was one that advanced and nurtured in others a worldview, which at its heart sought to divide and to equate national security with regime survival. Thus, the fate of the German people was depicted as being irrevocably linked to that of the Third Reich; if National Socialism were to go under, so too would – and should – Germany. And, as an important corollary, those who opposed National Socialism should be viewed equally as opponents of the German people. By drawing such a tight link between the fortunes of party and state, the Nazi regime attempted to provide not only for itself but also for the nation as a whole, a rationale for the ruthless suppression of its enemies.

And, indeed, there were many such 'enemies', for the ideology and practice of National Socialism not only cut across the core values and beliefs of many elements of German and Austrian society but invited them in response to either submit to or challenge its authority. Despite the personal dangers involved, a remarkable number chose to resist. The aim of this chapter is to capture and record that bravery. It does so by providing a succinct, yet comprehensive, overview of the myriad groups and organisations of differing political, religious and ethnic persuasions that took the fateful decision to resist the regime. The chapter identifies precisely when, where and with what frequency each of these groups carried out their serious resistance.

It is worth noting at the outset that answering such questions about resistance groups in a systematic manner is not as straightforward as one might think. It is a task vexed by a fundamental problem: How does one accurately classify resistance

W. Geerling, G. Magee, *Quantifying Resistance*, Studies in Economic History,
DOI 10.1007/978-981-10-6008-3_5

groups? As discussed in Chaps. 1 and 2, no commonly accepted approach currently exists. In fact, on closer perusal, the existing literature offers up to researchers a bewildering array of competing and contradictory classificatory systems, some of which appear not to even have had their roots in the historical analysis of resistance movements. As a result, many of the available schemes are either incomplete or conflate the political basis of resistance with features of its form, goal and location, problems which effectively render cross-study comparisons of specific resistance groups nigh impossible.

In this chapter, we avoid some of these pitfalls by following the classificatory approach adopted by an important source on resistance, the *Widerstand als Hochverrat* database. This framework is grounded in an understanding of serious political resistance as defined by the regime itself and demonstrated by its willingness to lay charges of treason and high treason in response (Zarusky and Mehringer 1998, p. 13). This classification's groupings incorporate known parties of resistance, groups based around significant resistance events, group designations created by the resistance literature itself and in a few instances major non-political organisations, of which resisters were active members. To avoid an unhelpful proliferation of categories, broader collective groups, where appropriate, are adopted to capture a wide variety of small resistance groups, which share common defining features. Resistance activities undertaken outside the structure of organised groups are simply classified as 'not organised'. The main advantage of this classificatory scheme, we believe, is that, unlike most others, it is historically accurate while also exhaustive and internally consistent in its coverage. In the following sections of this chapter, we use it as a framework for the accurate mapping of the activities of different resistance groups in Germany, Austria and outside the Reich.

5.1 Germany

Over the course of its life, the Nazi regime confronted serious resistance to its rule within Germany from a wide variety of sources. As Table 5.1 demonstrates, these included parties, groups and organisations from right across the political spectrum, each with their own varying motivations, objectives and ideological or theological underpinnings. It also included significant numbers of individuals acting independently of any formal organisation. The overwhelming majority of serious resistance, however, emanated from the left.[1] The Communist Party of Germany (*Kommunistische Partei Deutschlands* or *KPD*), together with a small number of its fellow Comintern-aligned Austrian counterparts in the Communist Party of Austria (*Kommunistische Partei Österreichs* or *KPÖ*), constituted the single largest group

[1] This finding is consistent with previous breakdowns of high treasonous resistance by groups. While drawing comparisons between studies is complicated by differing terminologies and definitions, both Marxen (1994, p. 40) and Form and Engelke (1996, p. 34), for example, identify a similar majority of resisters from left-wing parties and organisations.

Table 5.1 Arrests for treason and high treason in Germany, 1933–1945, by resistance group or organisation

	KPD/KPÖ	Trotskyist	*KPDO*	Other communist splinters	Communist help	*SPD*	*SAP*	Other socialists	Anarchist	Trade union	Youth
1933	418	5	3	3	0	11	7	0	1	0	0
1934	165	0	5	0	0	15	7	0	2	0	0
1935	322	6	1	0	7	82	14	0	5	0	0
1936	97	7	2	2	0	39	6	6	9	0	0
1937	123	2	9	20	6	25	1	4	6	0	0
1938	49	6	14	20	1	31	4	8	0	2	0
1939	92	0	0	0	0	4	1	0	0	0	0
1940	45	1	0	0	1	2	1	3	0	4	0
1941	89	0	0	0	6	3	0	1	1	0	0
1942	176	0	0	0	0	1	0	15	0	0	23
1943	178	0	0	0	0	7	0	26	0	0	0
1944	237	0	0	0	0	1	0	22	0	0	0
1945	0	0	0	0	0	0	0	2	0	0	0
Total	1991	27	34	45	21	221	41	87	24	6	23

	Black Front	National conservative	German Catholic	*Bündische Jugend*	German separatists	20 July	Austrian CCL	*Meuten*	White Rose	Not organised
1933	0	0	0	0	0	0	0	0	0	3
1934	14	0	0	0	0	0	0	0	0	1
1935	29	0	1	9	0	0	0	0	0	3
1936	11	2	7	2	0	0	0	0	0	4
1937	2	1	0	1	0	0	0	0	0	0
1938	12	2	0	0	0	0	0	8	0	4

(continued)

Table 5.1 (continued)

	Black Front	National conservative	German Catholic	*Bündische Jugend*	German separatists	20 July	Austrian CCL	*Meuten*	White Rose	Not organised
1939	9	5	0	8	9	0	9	0	0	2
1940	0	1	0	2	1	0	0	0	0	3
1941	0	0	0	0	0	0	0	0	0	7
1942	0	0	19	0	0	0	0	0	0	19
1943	0	0	1	0	0	0	0	0	25	56
1944	0	0	0	0	0	75	0	0	0	28
1945	0	0	0	0	0	1	0	0	0	0
Total	77	11	28	22	10	76	9	8	25	130

charged by the state with treason and high treason in Germany, a position it maintained even after it had abandoned open confrontation, gone underground and initiated a cell-based strategy of resistance.[2] Moreover, it should be noted that the *KPD*'s quantitative dominance held in every year of National Socialist rule, including 1939 and 1940, the key years of the German-Soviet Nonaggression Pact. Communist arrests in those years, did not cease. Indeed, the numbers of its members arrested for treason and high treason in 1940 differed little from that of the last pre-Pact year of 1938. Table 5.1 reminds us that, allowing for year-on-year fluctuations, the number of communist arrests had been in fact – like charges of treason and high treason in general in Germany – trending downwards since 1936. Further, it tells us that, in pure numbers alone, the years 1939 and 1940 were not that different from 1936 to 1938. Such figures and trends need to be borne in mind when assessing the impact of the non-aggression pact on communist resistance activities. Of course, following the invasion of the Soviet Union in June 1941, *KPD* arrests did renew their growth, reaching their wartime peak in 1944. Yet, even in that year of heavy resistance, the volume of *KPD* members charged with treason and high treason paled in comparison to that experienced during the two great pre-war waves of arrests of 1933 and 1935, when the fledgling Nazi regime ruthlessly crushed its main Weimar-era opponents. To put the scale of these pivotal events in perspective, the number of *KPD* arrests in those two pre-war years alone significantly exceeded the total number of its arrests for the entire period between the beginning of Operation Barbarossa and the end of the war.

SPD (*Sozialdemokratische Partei Deutschlands* or Social Democratic Party of Germany) arrests in the early years of the regime were also significant. By 1939, however, the party's leadership, by then operating in exile out of Paris, acknowledged Hitler's firm grip on power and called a halt to all active resistance within Germany, including the distribution of anti-regime pamphlets and party newspapers, in the belief that any further efforts along these lines would only prove futile and wasteful of human life. Thereafter, the focus of *SPD* activity for the remainder of the regime's existence shifted to the monitoring and reporting on popular opinion in Germany (McDonough 2001, pp. 3–4).[3]

In comparison to the *KPD* and *SPD*, the volume of arrests on the non-left side of German politics can only be described as minimal. Moreover, most non-left resistance was relatively small scale and transient. Indeed, in the pre-war years, the majority of those arrested were decidedly on the right: either members of the Black Front (*Schwarze Front*), a radical anti-capitalist political group formed by Otto Strasser after his expulsion from the Nazi Party in 1930, or part of the Buendische Youth (*Bündische Jugend*) movement, a loosely assembled association of independent German hiking groups, which espoused a traditionalist, nationalistic and anti-

[2] On the resistance of the KPD, see Weitz (1997, pp. 281–305), Merson (1985), Peukert (1980) and Holtmann (2010).

[3] For an overview of *SPD* resistance across the entire Nazi period, see Mehringer (2004, pp. 56–78).

democratic worldview.[4] In stark contrast, the most significant event for mainstream German nationalism or conservativism, the failed attempt on the life of Adolf Hitler on 20 July 1944, had to wait until the latter stages of the war to occur.[5]

More persistent and quantitatively important to serious resistance than the right of German politics, at least in terms of arrest for treason and high treason, were individuals and groups who had no obvious ties to organised political or religious organisations. While present across the entire Nazi period, such unorganised, unaligned – and, yet, still deemed by the regime itself as serious – resistance grew ever more common after 1942, as despair about the nation's trajectory began to weigh ever more heavily on the consciences of all, including even those not formally engaged in oppositional politics. Typical of this independently minded resistance was the White Rose group, a band of pacifist students and intellectuals based at the University of Munich, which, between June 1942 and the arrest of its core members in February 1943, produced and distributed pamphlets decrying the wanton crimes and brutality of the regime.[6] Similarly, the anti-regime and anti-authority activities of the eight members of the Leipzig *Meuten*, a teenage street gang rounded up in 1938, and of those belonging to three youth groups led by Herbert Baum, Helmuth Hübener and Walter Klingenbeck detained in 1942, reflected the fact that the rejection of, and resistance to, Nazi control and regimentation could extend to the very young indeed.[7] Far from being discounted as youthful exuberances, though, these challenges to its authority drew a harsh response from the Nazi state and, tragically, for 31 German teenagers and young adults, led to charges of high treason, for which 16 were ultimately to be executed.

As mentioned previously, comparing the figures presented in Table 5.1 with the political distribution of arrests for treason and high treason found by other researchers is complicated by the incompatibility of the classificatory schemes employed by different studies. However, by recasting the data behind Table 5.1 into a two-part categorisation utilised by Schlüter, namely, left-wing and remaining organisations, a comparison between his and our samples can be made, albeit at a very high level of aggregation. This comparison is presented in Table 5.2. It shows a very close parallel between the two studies in terms of the share of the left-wing among all resisters. Furthermore, it is noteworthy that, within the broad left-wing classifica-

---

[4]For a short overview of the Black Front, see Steinbach and Tuchel eds. (1998, p. 183); for the Buendsiche Youth, see Steinbach and Tuchel eds. (1998, p. 38).

[5]For the 20 July plot, see Ueberschär (2004, pp. 489–505) and Walle (2004, pp. 505–521). For an overview of conservative resistance, see Klause (2004, pp. 185–201).

[6]The standard biography of the White Rose comes from Inge Scholl; see Scholl (1983). For a more recent account, see Sturms (2013).

[7]Youth groups differed from teenage street gangs (*Meuten*) in that, unlike *Meuten*, they were not involved in fighting, petty crime or delinquency but had banded together typically because of a shared rejection of either the war, the actions of the regime or both. Each group tended to have a natural leader, whose name historians have used to identify the group. They consisted primarily of juveniles or, in the case of Baum's group, very young adults with little political expertise or direction. For an overview of the Leipzig *Meuten*, see Lange (2010). For Hübener, see Holmes and Keele (1995). For Klingenbeck, see Gleißner (2004). For Baum, see Benz (2014, pp. 48–51).

Table 5.2 Left-wing arrests for treason and high treason in Germany (Schlüter, p. 140)

	Geerling and Magee	Schlüter
Left-wing	85.6	88.0
Remaining organisations	14.4	12.0

Note: Schlüter's figures are derived from a sample of 308 People's Court cases

tion, Schlüter (1995, p. 140) attributed 69.1% of arrests to communist party and communist party-affiliated organisations, whereas our corresponding figure was an almost identical 69.0%. Indeed, the only significant difference between the two samples is how the remaining left-wing arrests have been divided between the *SPD* and other leftist organisations and parties. Schlüter (1995, p. 140) reports a higher *SPD* figure than we do (13.3% compared to 7.6%), a disparity which appears to have resulted from his different attributions of certain smaller left-wing groups. More importantly, though, the combined total of *SPD* and other left-wing arrests is similar in both studies (18.8% and 16.6%, respectively).

Looking more closely at the timing and relative importance of the resistance activities of the different left-wing groups, Fig. 5.1 not only confirms the overall preponderance of *KPD* resisters but also reveals what appears to be the regime's sequenced approach to the dismantling and destruction of serious left-wing resistance in the early pre-war period. Between 1933 and 1935, its primary focus was clearly on denuding the capability of the *KPD* to resist, and then the *SPD* and Socialist Workers Party of Germany (*Sozialistische Arbeiterpartei Deutschlands* or *SAP*) were steadily brought into frame in 1935 and 1936, followed finally by other smaller, presumably less regime-challenging, parties and organisations such as the anarchist-syndicalist Free Workers Union of Germany (*Freie Arbeiter-Union Deutschlands* or *FAUD*), the Communist Party of Germany (Opposition) (*Kommunistische Partei-Opposition* or *KPDO*) and a host of other small-scale socialist and communist parties not affiliated to the Comintern. Figure 5.1 also shows that, during the war, the *KPD* regained its overwhelming quantitative dominance of serious left-wing resistance, albeit with a growing competition, especially from mid-1943, from socialist groups, such as the Anti-Fascist Workers Group of Central Germany (*Antifaschistische Arbeitergruppe Mitteldeutschlands* or *AAM*) and the European Union (*Europäische Union*).

A similar analysis for right-wing groups is presented in Fig. 5.2. It highlights that the activity and arrests of non-left-wing resistance groups, with the exception of the Black Front, tended to occur in a series of sharp, pronounced bursts of fairly short duration. It was a pattern that in large part reflected the shallower organisational depth, narrower cohesion and greater issue- or event-orientated motivation of many of these groups. Waves of national conservative arrests occurred in the first half of 1937 and then again between July 1939 and June 1940. Arrests of Catholic groups spiked in the first half of 1936 triggered by the taking into custody of the group associated with Joseph Rossaint and then more prominently in 1942 and early 1943 with the detention of Johannes Prassek and the other Luebeck martyrs who had expressed publically their moral outrage at a number of actions of the regime includ-

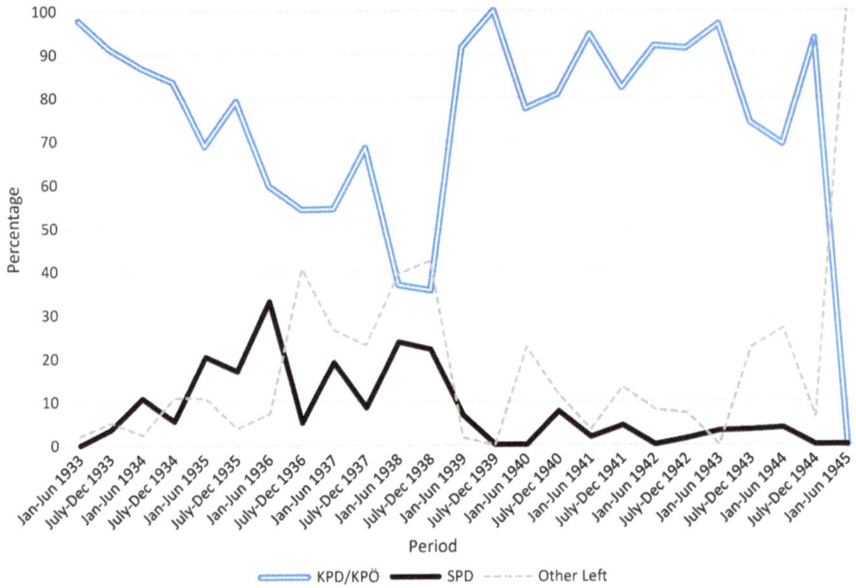

Fig. 5.1 Left-wing arrests in Germany by main parties and organisations, 1933–1945 (in percentages)

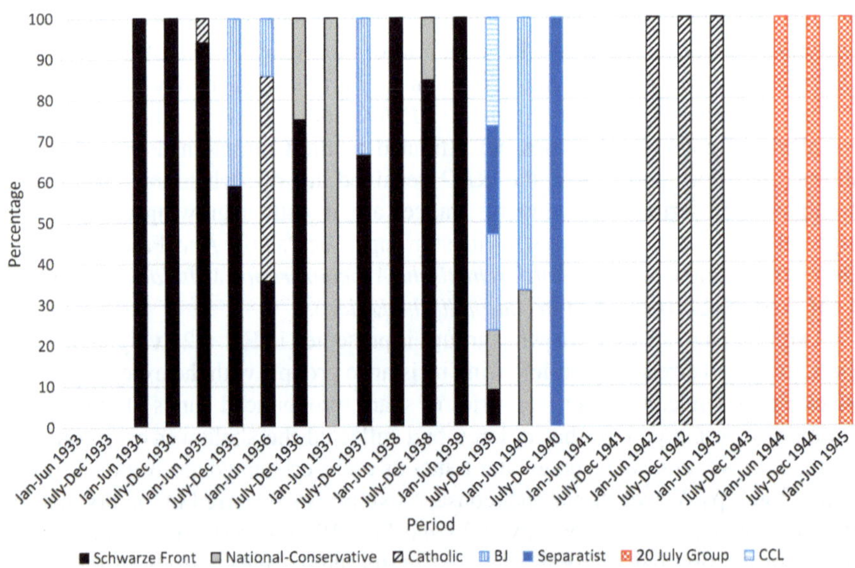

Fig. 5.2 Right-wing arrests in Germany by main parties and groups, 1933–1945 (in percentages)

ing its T4 euthanasia programme targeting people with mental illness or major disabilities, forced conscription into the *Wehrmacht*, persecution of religious orders and the closing of Catholic institutions.[8] A group of nine separatists led by Baron Adolf von Harnier, hoping to restore an independent Bavarian monarchy, were arrested in August 1939.[9] The largest non-left burst of activity of all, however, was the 20 July plot, which, in itself, accounted for around 20% of all arrests for high treason and treason in 1944 (Ueberschär 2004, pp. 489–505; Walle 2004, pp. 505–21).

Table 5.3 provides a snapshot of characteristics and experiences prevalent among members of different resistance groups. Coming from the numerically largest group, the *KPD*-aligned resister in many ways shaped the norm. Typically, he or she was a German in their mid-30s with a limited education that rarely went beyond the primary school level. One in three was either unemployed or not in the labour force, but, if employed, they were more than likely to have been engaged in a skilled or unskilled blue-collar occupation.[10] One feature that distinguished the *KPD* resister from all others, apart from their involvement in communist organisations, was the greater likelihood that they had had a previous conviction, especially for political crimes.

---

[8] Joseph Rossaint, a chaplain, left the Centre Party (*Zentrumspartei*) in 1933 after it voted in favour of the enabling act. He subsequently re-established – illegally – the disbanded Catholic Youth Organisation and put it in contact with the similarly banned Communist Youth Association of Germany (*Kommunistische Jugendverband Deutschlands* or *KJVD*). In 1935, both organisations produced and distributed a pamphlet critical of the reintroduction of compulsory military service in Germany. In January 1936, Rossaint and six others were arrested for these activities; see Jahnke and Rossaint (2002). Johannes Prassek and three other priests were arrested in 1942 for using their sermons to criticise openly the Nazi regime and to point out to their parishioners the inherently irreconcilable contradictions that existed between Catholicism and Nazi ideology. Together with 15 of their supporters, they also copied and distributed the sermons of Bishop August von Galen of Munster, who had led the church's protest against state-sponsored euthanasia. They were also accused of uttering subversive remarks about the war on a number of occasions; see Pelke (1961).

[9] Not too much should be read into the column for the second half of 1940. In that period, there was only one non-left arrest: an individual who had been campaigning for the independence of Alemannia, a historic tribal kingdom that extended across modern-day French Alsace, German Baden and Swabia, German-speaking Switzerland and Austrian Vorarlberg. He was arrested and charged with treason, for operating an illegal organisation, separatism and distributing a series of separatist leaflets in hotels, railways stations and restaurants across south and south-west Germany and Alsace in December 1940.

[10] This description of a usual *KPD* resister chimes with what we know about *KPD* members more generally. The *KPD* was the party of the unemployed, and a clear majority of its members – according to Weitz (1997, pp. 159, 245), 68.1% in 1927 – were skilled or unskilled workers. Among resisters, 52.5% were blue collar. The difference with the 1927 share of members is almost entirely attributable to the large proportion of party cadre among resisters (13.4% compared to 1.6% of the party membership in 1927). Not surprisingly, party cadres were more in the firing line of both resistance and Nazi persecution than the rank and file member of the party. If we exclude party employees from the breakdown, the share of skilled and unskilled blue-collar members in both samples converges dramatically: 66.4% of non-cadre resisters compared to 69.2 of non-cadre members in 1927.

Table 5.3 Personal experiences and characteristics of defendants arrested for activities in Germany, by resistance group (in percentages)

		KPD/KPÖ	Non-*KPD* communist	*SPD*	Other left	Conservatives and nationalists	Black Front	Not organised	Overall
Education and training									
Highest educational attainment	Primary	86.4	67.7	86.0	69.6	35.1	53.2	84.6	81.5
	High school medium	3.1	7.1	0.5	5.1	1.0	7.8	2.3	3.3
	High school advanced	4.2	9.4	6.3	4.4	9.3	15.6	5.4	5.4
	Tertiary	6.2	15.7	7.2	20.9	54.6	23.4	7.7	9.9
Vocational training	Completed or in process	59.0	52.0	51.6	53.8	13.4	20.8	56.9	54.9
Citizenship and ethnicity									
Citizenship	German	92.7	92.9	91.4	82.9	100.0	94.8	95.4	92.4
	Austrian	2.8	4.7	0.9	7.6	0.0	3.9	0.0	2.8
	Foreign	2.1	1.6	5.4	9.5	0.0	1.3	2.3	2.7
	Stateless	2.5	0.8	2.3	0.0	0.0	0.0	2.3	2.1
Ethnicity	Partial Jewish ancestry	2.2	6.3	2.7	5.1	3.1	2.6	1.5	2.6
Occupation									
	Student	0.9	0.0	0.9	1.9	0.0	2.6	0.0	1.0
	No particular occupation	4.9	1.6	4.5	3.8	3.1	5.2	9.2	4.8
	Blue collar (skilled and unskilled)	57.5	52.0	57.5	46.2	6.2	23.4	54.6	53.7
	White collar (excluding high-ranked state officials)	6.4	14.2	8.1	12.0	4.1	9.1	6.9	7.3

	Professionals (excluding military)	3.4	9.4	3.2	12.0	17.5	7.8	3.8	4.7
	Business or property owner	1.5	1.6	1.4	0.6	1.0	6.5	4.6	1.7
	High-ranked state official	0.2	0.0	0.0	0.6	26.8	0.0	1.5	1.1
	Engaged in full-time political work	13.4	0.0	5.0	1.9	6.2	1.3	0.8	10.2
	Enlisted soldiers or non-commissioned military officer	0.2	0.0	0.5	0.6	0.0	0.0	0.0	0.2
	Commissioned military officer	0.1	0.0	0.0	0.0	19.6	0.0	0.0	0.7
	Service worker	11.7	21.3	19.0	20.3	15.5	44.2	18.5	14.5
Employment status									
	Employed	67.4	70.9	66.5	79.1	97.9	77.9	76.9	69.9
	Unemployed	19.5	25.2	22.2	11.4	0.0	14.3	4.6	18.0
	Not in the labour force	13.1	3.9	11.3	9.5	2.1	7.8	18.5	12.1

(continued)

Table 5.3 (continued)

		KPD/KPÖ	Non-*KPD* communist	*SPD*	Other left	Conservatives and nationalists	Black Front	Not organised	Overall
Age									
	Mean age at arrest	36	34	37	36	52	36	45	37
Experiences									
Personal and familial	Chronic health problem	9.8	5.5	6.3	7.6	7.2	10.4	30.0	10.0
	Spent time outside the Reich (in a private capacity)	14.1	14.2	16.7	22.8	22.7	29.9	11.5	15.4
	Engaged or married	60.9	50.4	67.4	56.3	86.6	59.7	67.7	61.8
	Widowed	2.4	0.8	2.3	3.2	3.1	0.0	3.8	2.3
	Divorced	6.3	5.5	2.3	5.7	3.1	7.8	6.2	5.8
	Has children	28.3	18.1	28.1	28.5	42.3	26.0	44.6	28.9
	Came of age during 1918–1919 Revolution	25.2	31.5	25.3	17.7	18.6	24.7	15.4	24.1
	Came of age during the hyperinflation	37.3	44.9	29.9	32.3	16.5	35.1	19.2	34.8
	Immediate family engaged in anti-regime activities	15.2	15.7	12.7	20.9	28.9	19.5	18.5	15.9
	Served in World War I	24.3	18.1	38.5	19.6	42.3	28.6	39.2	26.1
	Served in World War II	2.0	0.0	0.0	8.2	25.8	0.0	3.1	2.9
	Family member killed during war	1.5	0.0	0.5	1.3	4.1	0.0	6.9	1.6
Religious	Motivated by strong religious convictions	3.0	3.1	2.3	3.8	32.0	13.0	4.6	4.6
	Roman Catholic	2.0	1.6	1.8	2.5	23.7	13.0	3.8	3.3
	Evangelical	1.0	1.6	0.5	1.3	8.3	0.0	0.8	1.2

Legal	Previous convictions	39.0	16.5	21.3	23.4	10.3	31.2	40.8	34.4
	Previous terms of imprisonment	34.0	12.6	15.8	18.4	9.3	29.9	37.7	29.7
	Previous major political convictions	13.5	3.9	1.8	5.1	3.1	3.9	11.5	10.9
	Previous minor political convictions:								
	Defamation	5.2	2.4	3.2	2.5	1.0	7.8	4.6	4.6
	Public disorder	5.0	0.8	0.0	3.2	0.0	1.3	1.5	3.9
	Resisting state authorities	3.5	0.8	0.9	3.2	1.0	0.0	1.5	2.8
	Illegal meetings and/or writings	5.6	2.4	1.4	3.2	1.0	0.0	3.1	4.5
	Malicious gossip	0.3	0.0	0.0	0.6	1.0	0.0	3.1	0.4
	Previous non-political convictions:								
	Property crimes	11.6	2.4	5.0	5.1	2.1	13.0	16.2	10.2
	Sexual offences	1.2	0.8	1.8	0.6	0.0	1.3	5.4	1.4
	Crimes of violence	7.3	2.4	4.1	5.1	1.0	6.5	7.7	6.4
	Other crimes	12.8	7.1	10.9	6.3	3.1	24.7	9.2	11.8

(continued)

Table 5.3 (continued)

		KPD/KPÖ	Non-*KPD* communist	*SPD*	Other left	Conservatives and nationalists	Black Front	Not organised	Overall
Organisational memberships (current or past)	Denied entry/expelled from Hitler Youth/ League of German Girls	0.3	0.8	0.0	0.6	1.0	0.0	1.5	0.5
	Non-political and non-religious	15.8	28.3	22.2	19.0	3.1	11.7	11.5	16.2
	Trade union	28.4	48.0	55.7	34.2	6.2	9.1	23.8	30.1
	Hiking group	1.1	7.1	0.9	3.8	5.2	2.6	0.8	2.2
	Protecting the Weimar Republic	1.4	3.1	31.7	7.6	4.1	1.3	2.3	4.3
	Free Corps/Association of Front Soldiers	3.8	0.8	4.1	2.5	20.6	24.7	4.6	4.8
	SA/*SS*	1.5	0.8	0.9	1.3	7.2	22.1	2.3	2.3
	Nazi Party (German or Austrian)	1.8	0	0.9	1.3	14.4	53.3	4.6	3.7
	Communist	71.9	61.4	5.0	17.7	0.0	3.9	30.0	9.1
	Social Democrat	13.8	23.6	84.2	44.3	19.6	2.6	22.3	21.6
	Socialist	9.6	11.8	6.3	34.8	4.1	0.0	6.9	4.5
	National conservative	1.2	1.6	0.5	3.2	45.4	14.3	2.3	27.3
	Centre Party	0.6	0.8	0.5	0.0	14.4	9.1	1.5	31.8
	Anarchist	0.4	0.0	0.0	15.2	0.0	0.0	0.0	1.1
	Trotskyist	0.1	17.3	0.0	0.0	0.0	0.0	0.0	0.8

Note: The category ‘Conservatives and nationalists’ consists of all national conservatives, German separatists and members of the 20 July group

Theodor Kühnlein from Nuremberg was fairly representative of resisters working for the *KPD*. After completing primary school, he had undertaken an apprenticeship as a baker but, being unable to find work in this field, was forced to take on a string of temporary unskilled jobs in factories and on farms and construction sites. He also experienced periods of protracted unemployment, which appear to have radicalised him. Married with two children, he joined the *KPD* and Revolutionary Trade Union Opposition (*Revolutionäre Gewerkschafts-Opposition* or *RGO*) in 1931 and worked as an election helper for the party in the last free elections of the Weimar era. Although he was formally expelled from the *KPD* in 1932 following a dispute with a local functionary, he remained closely engaged with the party, helping to keep a number of its cells in the region active in the pre-war years. Arrested in January 1939 at the age of 37, he was convicted and sentenced by the People's Court to 4 years in a penitentiary for membership of a banned organisation (Anklage 9 J 355/39 g, pp. 1–4; Urteil 2H 9/40, pp. 1–4).

Relative to the *KPD* member, other leftist resisters were more likely to have had a high school or tertiary education, to have been partially Jewish or of foreign extraction and to have been members at some point of a trade union and a social democratic organisation. Reflecting both the less structured nature of many of their organisations and their shorter histories, fewer non-*KPD* left resisters were engaged as political cadres or had had previous political convictions. Not surprisingly, affiliations with leftist views and opinions not aligned with the prevailing official interpretation of Stalinism, such as Trotskyism (among non-*KPD* communists) and anarchism (among other left), were more acceptable to, and common among, the non-*KPD* left.

The experiences of Walter Uhlmann, a trained mechanic with a primary school education, were typical of many on the non-*KPD* left incarcerated for their resistance. After working for short periods in Leipzig, Hannover and Cologne, he moved to Berlin in 1929 at the age of 25 in search of work but, apart from occasional temporary jobs, remained basically unemployed till 1934. Attracted to communism since his youth, he joined the Free Socialist Youth (*Freies Sozialistische Jugend*) in 1919 and ultimately the *KJVD* in 1923. He worked for the district leadership of the *KJVD* in Rhineland but was expelled from that organisation in 1928 over his disagreement with the *KPD*'s tactics, especially in relation to the creation of the *RGO*. He joined the Communist Party of Germany-Opposition (*Kommunistische Partei-Opposition* or *KPDO*) in 1930 and became a trade union spokesman for metal workers in Berlin. Between 1934 and his arrest in February 1937, he was a member of the Berlin Committee of the *KPDO* and was pivotal in organising an illegal underground network of left-leaning metal workers in the capital (Anklage 17 J 152/37, pp. 1–3; 8–9; Urteil 2H 51/37, pp. 1–2; 8–9).

Table 5.3 reveals that two other types of resisters – those belonging to the Black Front and non-aligned individuals – shared very much in common with left-wing resisters. There were differences, of course. Apart from their obvious history of ties to the Nazi Party and its ancillary organisations, the Black Front resister tended to be better educated, with nearly one in four having had a tertiary education, and to have been more inclined to be motivated by their Roman Catholicism. By contrast, the individual resister was on average about 10 years older than their *KPD* counterpart and more prone to chronic illness, to be outside the labour force, to have had children and to have served in World War I.

One such unorganised and unaligned resister was Willi Kranz, at the time of his arrest in March 1943, a married stoker with three children residing in Dessau in Saxony-Anhalt. Born in 1897, he received only a basic education and served in the army during World War I. He received the Iron Cross (second class) for gallantry. After the war, he worked in a variety of blue-collar jobs and also did some time in jail for theft. Although sympathetic to the *SPD* and a member of the Metal Workers Trade Union (*Deutscher Metallarbeiter-Verband* or *DMV*), he never joined any political organisation. His troubles with the regime began in early 1943, when he and five other disillusioned co-workers at a large armaments factory in Dessau-Rosslau openly and repeatedly made subversive and defeatist remarks to colleagues about the parlous state of the war, how life would soon be better under the Soviets and how, when that time came, they should all inform on who had been the Nazis in their midst. They were to prove fatal utterances. For his subversive and high treasonous activities, Kranz was sentenced to death by the People's Court (Anklage 5 J 120/43, pp. 1–3; Urteil 2H 141/43, pp. 1–4).

Perhaps the types of resister that tended to be the biggest outliers in terms of characteristics and experiences were those who came from the conservative and nationalist side of German politics. All ethnically German, compared to leftist resisters, they were considerably older (by, on average, 15 years), more often than not university-educated veterans of either or both World Wars, and more likely to have had a history in a Free Corps (*Freikorps*) or the Association of Front Soldiers (*Stahlhelm*). They had had fewer prior run-ins with the law and, at the time of their arrest, were typically gainfully employed as professionals, senior bureaucrats or military officers. In their personal lives, they were almost all married or engaged and were more likely to have had children but, reflecting their higher socioeconomic circumstances, less burdened by chronic health problems. Prior to arrest, they were more inclined to have been influenced by strong religious convictions and have had their political views expressed through memberships of either a national conservative organisation, the Centre Party or, indeed, the Nazi Party.

Characteristic of many conservatively inclined resisters was Walter Knopf, the married son of a Catholic customs inspector, born in Hillersleben in Saxony-Anhalt in 1885. After studying linguistics at university, he took and passed the entry exam for high school teachers then did a year of military service in 1912. Afterwards, Knopf worked as an assistant teacher at a Catholic high school in Beuthen, Silesia. At the outbreak of war, he joined the army again and was wounded in September 1914, returning to the front only in September 1917, this time as a sergeant. In July 1918, he received further wounds, was released from the army and eventually resumed his teaching career in Beuthen. There he remained until August 1933, when, under Section 4 of the Law for the Restoration of the Professional Civil

Service (*Das Gesetz zur Wiederherstellung des Berufsbeamtentums*), Knopf was dismissed from his position for his negative attitudes to National Socialism (Reichsministerium des Innern, *Reichsgesetzblatt* (RGBl), 7 April 1933, Part I, p. 175). Knopf believed that National Socialism was incompatible with the teachings of Christianity and accordingly tried to prevent his son from joining the Hitler Youth, relenting only when it was made clear that such action would render his son ineligible to take exams. Drawn to politics after the war, he had joined the Centre Party in 1918 but left 10 years later feeling that the party had become too dependent on the right wing. He joined the Christian Social Reich Party (*Christlich-Soziale Reichspartei*), a small Christian pacifist party, and then later the German Freedom Party (*Deutsche Freiheitspartei* or *DFP*), a conservative opposition group founded by German exiles in Paris in 1937. Knopf was involved in the illegal importation and dissemination of high treasonous materials, such as 'The German in Poland' (*Der Deutsche in Polen*), a newspaper, produced in Katowice before World War II, which promoted Christian political values and criticised Nazi policy and ideology. Arrested in October 1939, Knopf was sentenced to life in a penitentiary (Urteil 2H 89/40, pp. 1–6).

Turning to the question of location, Table 5.4 reports the regional distribution of serious resistance activities by the main parties and groups in Germany. While the larger established parties, such as the *KPD* and *SPD*, were operational across almost all parts of Germany, other groups had tighter regional foci. That said, Table 5.4 illustrates yet again the vital importance of Berlin as a major – if not the major – location of resistance for all types of resister other than the Buendische Youth (*Bündische Jugend*), whose main regions of activity were Schleswig-Holstein and Hamburg, Lower Saxony and Bremen, Westphalia and the Rhineland (North and South). Saxony also emerges as a very important location of resistance for all left-wing groups as well as the Black Front and the 20 July group, while, outside Berlin, the loci of national conservative resistance were Silesia, Lower Saxony and Bremen, Schleswig-Holstein and Hamburg and Rhineland North. As for unorganised resisters, their presence was felt in most parts of the country with large clusters occurring in Westphalia, Berlin and Hessen. Nonparametric statistical analysis of the regional distribution of activity of the two largest groups of German resisters, the *KPD* and *SPD*, shows that the share of resistance that each exhibited in a region did not reflect the proportion of their supporters nationally that resided in that region, as gauged by the number of votes each had attracted there in the November 1932 Reichstag Election. This finding suggests that one did not simply resist where one lived; the choice of location was driven, it seems, by other, more circumstantial and strategic considerations.[11]

---

[11] For the sake of readability, the chi-squared statistical tests underpinning this finding have not been reproduced here. Further details of all tests are available from the authors upon request.

Table 5.4 Regional pattern of activity by main resistance groups in Germany, 1933–1945 (in percentages)

	KPD/KPÖ	*SPD*	National conservative	20 July group	Black Front	Non-*KPD*/ KPÖ communist	Other left	Buendische Youth	Not organised
Baden-Wuerttemberg	5/4	8.1	0.0	3.3	5.5	5.7	5.4	9.5	3.8
Bavaria North West	1.0	4.2	0.0	6.0	6.7	1.6	4.1	0.0	5.0
Bavaria South East	2.2	0.8	0.0	6.6	3.0	4.9	10.4	0.0	6.9
Brandenburg I (Berlin)	32.0	15.1	45.5	47.7	19.4	24.8	21.6	6.3	11.9
Brandenburg II	1.8	2.1	0.0	3.3	0.0	0.4	6.2	0.0	0.0
East Prussia	1.0	0.0	0.0	2.0	0.0	0.0	0.0	0.0	1.9
Hessen	4.3	3.4	0.0	2.0	1.8	2.4	5.0	1.6	10.1
Lower Saxony and Bremen	4.4	6.0	27.3	5.3	7.3	7.7	5.0	15.9	5.7
Pomerania-Mecklenburg	1.3	1.3	0.0	0.0	0.6	0.0	10.0	0.0	0.0
Rhineland North	10.5	10.4	27.3	0.0	6.1	2.8	5.0	17.5	7.5
Rhineland South	3.0	7.3	0.0	0.0	6.1	5.7	5.0	15.9	3.8
Saarland	2.2	0.5	0.0	2.0	4.8	2.4	0.0	0.0	3.8
Saxony	12.0	18.0	0.0	16.6	16.4	14.2	12.9	1.6	6.3
Saxony-Anhalt and Thuringia	4.3	9.4	0.0	0.0	9.1	13.0	2.1	0.0	5.7
Schleswig-Holstein and Hamburg	8.6	8.1	36.4	0.7	4.8	8.1	1.2	15.9	2.5
Silesia	2.2	1.0	45.5	3.3	8.5	4.9	5.0	0.0	2.5
Westphalia	3.9	4.2	0.0	1.3	0.0	1.2	1.2	15.9	22.6

5.2 Austria

Austria was the first country to be invaded by the Nazis. Yet unlike other victims in Hitler's quest for 'living space' (*Lebensraum*), with the exception of the Sudetenland and the Baltic territories, *Wehrmacht* soldiers were welcomed in Austrian cities as liberators rather than conquerors. Following the annexation of Austria (*Anschluß*) of 12 March 1938, the newly installed National Socialist government quickly assumed the status of a legal and legitimate power. Resisters in Austria could not count on the same support as in other occupied countries where collaborators were isolated and resistance was viewed as part of a war of liberation. There was no Austrian government in exile able to offer financial and military support to groups inside Austria. Those who became resisters did so without the support or assistance of the majority of Austrians who regarded them as traitors, creating a climate of repression and fear where denouncement by supporters of the regime was a continual threat.[12] Thus, the conditions under which resistance operated in Austria and Germany were similar, as, too, were the tactics employed there by the *Gestapo* and judiciary: a mix of prohibition, arrest, terror, lengthy sentences of imprisonment and execution.

Nor should the role of ordinary Austrians in Nazi crimes be overlooked. The argument postulated by the outgoing Austrian Chancellor, Kurt Schuschnigg, immediately following the events of March 1938 – that Austria had been the first victim of Nazi aggression – has a hollow ring to it when one considers the participation of ordinary Austrians in the final solution, military expansion and the racist war in the east.[13] Besides the prominent culprits, there were thousands of faceless people who saw in the greater German Reich an opportunity for career advancement.[14]

---

[12] The *Anschluß*, of course, despite its initial popularity, did have dramatic and increasingly wearing consequences for the lives of all Austrians. They were henceforth required to fulfil the same totalitarian claims as Germans had since 1933. As nationals of the Reich, they could expect the same protection – or lack thereof – before the law as their German counterparts too. All of their political and religious organisations outside state-sanctioned bodies were, for example, disbanded. In addition, according to the Decree for Youth Service (*Jugenddienstverordnung*) of March 1939, all Austrian male youth were now forced to perform unpaid work service, undergo gruelling physical training and political indoctrination as part of their mandatory attendance in the Hitler Youth and ultimately volunteer or be conscripted into the *Wehrmacht*; see Hagspiel (1995, p. 126).

[13] Approximately 40% of the personnel and most of the commandants of the death camps and 80% of Adolf Eichmann's staff were recruited from Austria. In addition to Eichmann, the man assigned the task of coordinating Jewish resettlement from Vienna in 1938–1939 and one of the leading architects of the Holocaust, many other prominent Nazis came from Austria: Baldur von Schirach, Reich Leader of Youth between 1933 and 1941 and Party leader for the district of Vienna afterwards; Arthur Seyss-Inquart, Chancellor of Austria after the union; and Ernst Kaltenbrunner, secretary of state for security purposes in Austria in 1938 and Chief of Police 1938–1942. See Pohanka (1997).

[14] After the *Anschluß* was proclaimed, there was the inevitable rush for Nazi Party membership. In Vienna, 274,000 new applicants were lodged by December 1938. Austrian Party members were allocated numbers from 6,100,000 to 6,900,000. Membership reached a peak of 693,007 in March 1943; see Jagschitz (1988, pp. 504–05). There were 11,560 Austrians in the SS in 1938. The figure rose to approximately 20,272 by 1941 and stayed relatively constant until the end of the war; see Jagschitz (1988, p. 506).

Still, not everybody looked forward to life under the swastika. For Jews, communists, social democrats, union leaders and other perceived enemies of the state, the *Anschluß* promised expulsion from the national community, followed by persecution, incarceration and often death.[15] Within days of Hitler's triumphant march into Vienna, the *SS*, police and local volunteers began clearing the streets of 'undesirables'. Anyone who could be a threat to the new regime was targeted: the usual political and racial enemies of the state and opponents of the *Anschluß* were given special priority. Before the plebiscite was held on 10 April 1938, over 70,000 functionaries of the 'fatherland' parties, socialists, communists and Jews had been arrested. Thousands more emigrated, went into hiding or abandoned activities perceived to be hostile to the state in favour of passive acceptance. It came as no great surprise when 99.7% of the population supported the plebiscite. Only 11,281 people voted against it (Stadler 1971, p. 184).[16]

Despite the cloak of legality surrounding the *Anschluß*, many Austrians saw, or came to see, the following 7 years as a period of foreign occupation and became involved in opposition and resistance as a consequence. Table 5.5 breaks down the incidence of resistance in Austria, which, between 1938 and 1945, was serious enough to be deemed treasonous or high treasonous by the regime and tried before the People's Court. It includes German groups active in Austria, most notably, the *KPD*, Black Front, the White Rose and a host of small German leftist organisations, which together accounted for 4.2% of all Austrian-related arrests. With respect to locally sourced resistance, Table 5.5 indicates that two parts of Austrian society offered the greatest challenge to the Nazis: the organised political parties of the working class and the nationalist, catholic and monarchist organisations of the bourgeoisie. The former, the significantly larger of the two, was splintered, having suffered from years of persecution prior to the *Anschluß*, and was to be easily targeted and removed by the *Gestapo*; the latter lacked experience in resistance, leadership and had no uniform ideological base. Table 5.6 indicates that our estimates of the relative size of working-class and catholic-conservative-legitimist resistance in Austria match closely with those presented by Bailer and Ungar (2013, pp. 114–15).[17]

The existence of socialist resistance organisations had its roots in the events of February 1934, when the Chancellor, Engelbert Dolfuß, in the aftermath of the 3-day Austrian civil war and the ensuing consolidation of his Fatherland Front gov-

---

[15] In the 7 years of Nazi terror that followed the *Anschluß*, 100,000 Austrians were arrested, 2700 resistance fighters were hanged, 32,000 died in concentration camps and prisons, 65,000 Jews were killed and 160,000 Jews were forced to emigrate; see Hartl (1995, pp. 79–80).

[16] Thousands of political and racial enemies of the state had been taken into 'protective custody', murdered or had emigrated and so were not represented in the vote. Nevertheless, this figure is a reflection of the popular support for the Nazis. For the number of arrests, see Hartl (1995, pp. 79–80). Neugebauer gives an estimate of 50,000–76,000 for the same period; see Neugebauer (1988a, p. 164).

[17] Neugebauer (2013, pp. 239–40) similarly finds that a clear majority of those appearing before the People's Court or the Higher Regional Courts in Vienna and Graz had, or had held, membership of a left-wing working-class political organisation.

Table 5.5 Arrests for treason and high treason for activities in Austria, 1938–1945, by resistance group or organisation

	KPÖ	*SPÖ*	*RSÖ*	CCL	Austrian communist help	Austrian Trotskyist	*Wiener Mischlingsliga*	Austrian freedom front	Slovenian Partisans	Not organised
1938	17	0	7	7	0	0	0	0	0	0
1939	23	0	0	38	0	0	0	0	0	1
1940	46	0	0	78	0	0	0	0	0	0
1941	272	0	0	8	23	0	0	0	0	2
1942	348	0	29	8	24	0	0	0	0	3
1943	175	0	0	45	40	7	0	0	0	26
1944	60	3	0	24	23	0	12	23	13	6
1945	0	0	0	0	0	0	0	0	0	0
Total	*941*	*3*	*36*	*208*	*110*	*7*	*12*	*23*	*13*	*38*

	Across party lines	Other Austrian left	Austrian youth	*KPD*	*KPDO*	*SPD*	Other German left	Black Front	White Rose
1938	0	0	0	2	1	0	0	8	0
1939	0	0	0	3	0	0	0	2	0
1940	0	7	0	1	0	1	0	0	0
1941	0	0	1	3	0	0	0	0	0
1942	0	1	3	12	0	0	4	0	0
1943	0	2	0	5	0	0	0	0	14
1944	18	0	0	0	0	0	6	0	0
1945	0	0	0	0	0	0	0	0	0
Total	*18*	*10*	*4*	*26*	*1*	*1*	*10*	*10*	*14*

Table 5.6 Working-class and conservative resistance in Austria (Bailer and Ungar 2013, pp. 114–15)

	Geerling and Magee	Bailer and Ungar
Working class	84.0	81.9
Catholic-Conservative-Legitimist	16.0	18.1

Note: Bailer and Ungar's figure relates to Austrian victims of political persecution in general. The table adopts Bailer and Ungar's terminology. The figures presented, thus, show two alternative breakdowns of those deemed 'politically' active on the left and right of resistance activity in Austria. Those not formally aligned with either a left- or right-wing political party or organisation are not included in either set of figures presented

ernment banned the Social Democratic Party of Austria (*Sozialdemokratische Partei Österreichs* or *SPÖ*) and trade unions, in effect forcing the Austrian workers' movement underground. Out of the various socialist resistance groups that emerged and continued on into the post-*Anschluß* era, the Revolutionary Socialists of Austria (*Revolutionäre Sozialisten Österreichs* or *RSÖ*) became the successor to the defeated *SPÖ* and displayed a willingness to sacrifice themselves in the resistance to the Dolfuß-Schuschnigg regimes (*DÖW* 1975, vol 2, p. 7). The *RSÖ* was no match for the *Gestapo*, however. The police knew most of the leading functionaries, had lists of former social democrats and with the help of espionage within the ranks of the movement was able to wipe out almost the entire political leadership shortly after the *Anschluß*.[18] In face of such losses, its central committee in March 1938 called off all activities, inducing many of its members who wished to continue the struggle to cross over to the *KPÖ* or other communist groups; it was not until 1942 that a renewed and concerted burst of *RSÖ* resistance had to be quelled again by the regime.

As Fig. 5.3 and Table 5.5 clearly illustrate, it was the *KPÖ* that provided the numerically greatest and most sustained challenge – both from the left and overall – to National Socialist domination in Austria. When the contribution of ancillary organisations like Communist Help is added to the picture, its overwhelmingly pre-eminent position in the Austrian-grown resistance movement is indubitable. The *KPÖ* had also been the only political organisation that had tried to rally all anti-fascist forces together to fight the common enemy – Nazism – before the *Anschluß*. After the Seventh World Congress of the Communist International, held in 1937, the *KPÖ* proposed the creation of a People's Front (*Volksfront*), where communists, socialists, farmers, intellectuals and Catholics would join together, similar to the French Popular Front in the fight against fascism (*DÖW* 1975, vol 1, p. 216; Neugebauer 2013, pp. 243–44). Post-1938, a curious mix of Marxism-Leninism and

[18] For a brief account of *SPÖ* and *RSÖ* resistance, see Hagspiel (1995, pp. 302–03) and Hormayr (2012). To uncover cells of resistance, the Gestapo used agent provocateurs, often previously arrested members of a particular group, to solicit further contacts and pave the way for future arrests. Hans Pav, a socialist, was one such double agent. Arrested in March 1938, he was released in order to meet with his socialist comrades and help rebuild the Revolutionary Socialist movement. The entire *RSÖ* central organisation was later arrested as a result of Pav's work. For other examples of espionage within the *SPÖ* and *RSÖ*, see Neugebauer (1988b, p. 540).

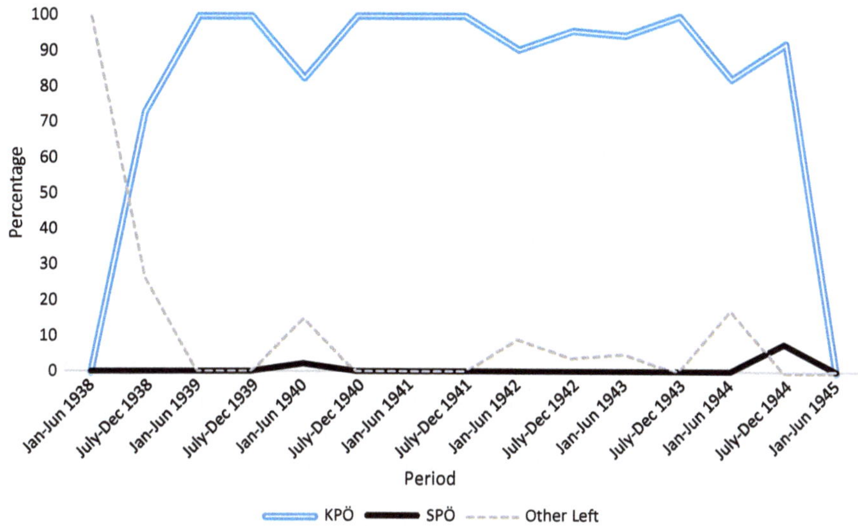

Fig. 5.3 Left-wing arrests in Austria by main parties and organisations, 1938–1945 (in percentages)

Austrian patriotism remained characteristic of its propaganda. This departure from the standard class struggle, proved in vain, however; the *KPÖ* – then as later – never came to be viewed as a serious alternative for most Austrians.[19]

The *KPÖ* suffered more persecution than any other resistance group. The Nazi agencies of terror sought to undermine the *KPÖ* through a systematic campaign of terror, arrest, infiltration and murder. Entire networks and cells were uncovered after the *Gestapo* arrested leading functionaries.[20] The centralised leadership structure of the *KPÖ* inadvertently left behind a trail of members, supporters and sympathisers for the *Gestapo* to round up.[21] In 1939, members of the Communist Youth Association of Austria (*Kommunistische Jugendverband Ősterreichs* or *KJVÖ*) were forced to fill the void left by the loss of experienced functionaries who had been murdered or arrested. The leadership structure was also reorganised to help combat the high casualty rate. A central committee from outside Austria took over leadership duties; following the lessons of its German counterpart, new party regulations imposed a maximum number of one hundred members in each regional branch, while smaller

[19] Neugebauer (1988b, pp. 541–42) and Hagspiel (1995, p. 305). Gestapo reports reveal that in the period March through December 1938, 5000 pamphlets, 127 leaflets and 1005 promotional brochures were discovered in Vienna and Lower Donau; see Botz (1988, p. 470). *DÖW* estimate that 90% of all anti-Nazi leaflets in Austria came from Communists; see Botz (1988, p. 472).

[20] Of 2800 cases tried before the Higher Regional Courts of Vienna, Graz and Klagenfurt, 80% of the 7500–8000 people tried were *KPÖ* members. A similar percentage holds for the People's Court; see Neugebauer (1988a, p. 178) and Neugebauer (1988b, p. 541).

[21] One such operation occurred in the autumn of 1939. The Gestapo smashed the leadership of the Viennese branch, arresting 206 people; see Botz (1988, p. 472).

cells of three to five agitators operated in either territorial or factory units (Hagspiel 1985, p. 305; Stadler 1971, p. 169). In theory, smaller decentralised units would permit greater flexibility in tactics and prevent mass arrests by limiting full knowledge of the party's hierarchical structure to only a handful of people. Of course, the prescribed rules for numbers in the cell were often overlooked; thus when caught, large numbers of people and complete groups were often apprehended (Konrad 1978, p. 23). Throughout the war, the *KPÖ* continued to agitate in factories, especially within the armament industry, and was undoubtedly a constant source of annoyance for the Nazis. However, never able to mobilise the working class or embrace bourgeois groups to present a united front against Nazism, it was never to be more than that.

In one of the earliest studies of Austrian resistance, Karl Stadler (1971, p. 156) argued that Austrians were 'motivated by political and social experiences, by moral and religious objections, and not by consideration of nationality'. Objections to National Socialism stemmed from a rejection of 'National Socialist policies, economic injustices, religious persecution—in short, to the manifestations of a particular ideology and not to Germany and the Germans' (Stadler 1971, p. 154). The political and social discontent fuelled a desire for national liberation among Austrians – as it did in France, Yugoslavia, Poland, the Netherlands and other European countries under Nazi rule – because the restrictions imposed by Nazi rule did not allow mechanisms of protest permissible in a democratic society. A campaign for the restoration of a free, independent and Catholic Austria attracted supporters of former Crown Prince Otto von Habsburg, separatists, nationalists and conservative opponents of the *Anschluß* and became a national crusade, drawing many ordinary, previously law-abiding citizens into loosely based circles of resistance to protest against the loss of Austrian independence. Such feelings coalesced around a dozen or so pro-independence groups that emerged during the early years of Nazi rule. In so far as a broad classification is possible, these groups which arose mainly after September 1939 were generally drawn from conservative or bourgeois circles and strove for the overthrow of the National Socialist leadership, the cessation of Austria from the Reich and the resurrection of an independent Austrian state along democratic lines (Wagner 1974, pp. 447–48). These goals were typically pursued through the distribution of leaflets, the re-establishment of contacts with emigrants and the collection of money for families of people imprisoned.

One such group was the Austrian Freedom Movement (*Österreichische Freiheitsbewegung* or *ÖFB*), formed in mid-1939, which derived its support from Catholic and middle-class elements (Neugebauer 2013, pp. 248–49). At its peak, this group centred around Roman Karl Scholz, Jakob Kastelic and Karl Lederer had in excess of two hundred members.[22] Other groups included the Anti-Fascist

[22] By April 1940, the Austrian Freedom Movement was chiefly comprised of a coalition between Lederer's original group of the same name and Kastelic's Great Austrian Freedom Movement (*Großösterreichische Freiheitsbewegung* or *GÖFB*), an organisation established in 1938 to advocate for an independent, democratic Austria under a Habsburg monarch. An example of the people who were attracted to Kastelic's movement was Florian (adopted religious name: Gebhard) Rath,

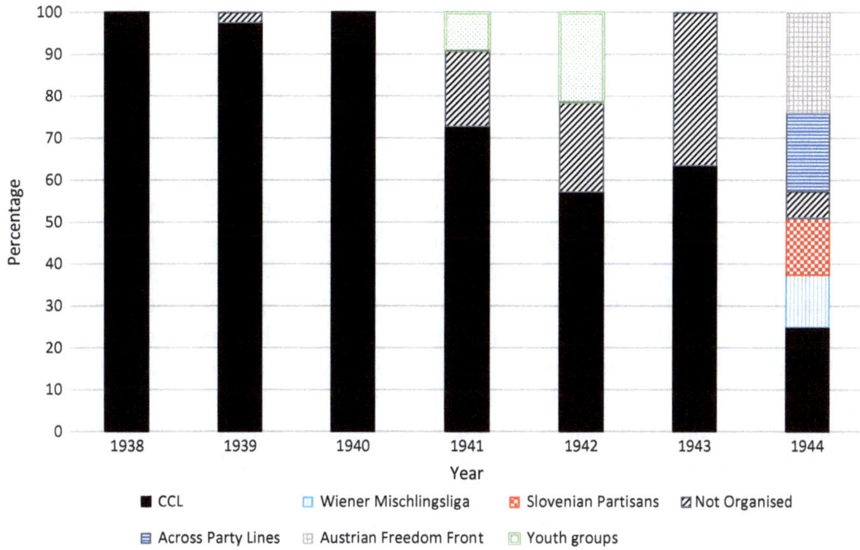

Fig. 5.4 Non-left-wing arrests in Austria by main parties and groups, 1938–1945 (in percentages)

Freedom Movement for Austria (*Antifaschistische Freiheitsbewegung Österreichs* or *AFÖ*) and the Austrian Underground Monarchist Front (*Illegale österreichische kaiserliche Front* or *IÖKF*), a group connected to the legitimist movement hoping to restore Habsburg rule (Botz 1988, p. 473; Stadler 1971, pp. 178–79). Together, groups such as these (labelled CCL in Table 5.5 and Fig. 5.4) constituted the largest non-left group of resisters in Austria and were most active in two periods: 1939–1940 and then in 1944. Given the diverse range of interests, origins, goals and motives encapsulated by the different groups involved, a lack of unity between the various circles of Catholic-Conservative-Legitimist resistance was inevitable. It was a weakness which Neugebauer (1989, p. 83) attributed to prevailing historical and cultural factors: 'a common root of resistance, which notwithstanding the existing political differences, was characteristic of other occupied countries, was, on account of the special 'national' situation of Austria, absent for a long time. It developed only towards the end of the war'.

As Figs. 5.3 and 5.4 neatly illustrate, until the end of 1943, almost all arrests for Austrian-based treason and high treason came from the two sources already described: the communist and catholic-conservative-legitimists movements. The

an ordained Cistercian priest from Gramastetten, Upper Austria, who obtained his doctorate in history from the University of Vienna in 1931 and until his arrest with Kastelic and two others in July 1940 worked as an archivist and librarian. He had apparently heard about the group through word of mouth and had decided to become involved in its activities early in 1940. For his association with a banned organisation and involvement in 'separatism', Rath was sentenced to 10 years in a penitentiary by the People's Court. He was 41 years old. Kastelic, the group's leader, was executed on 2 August 1944; see Anklage 7(8)J 203/41, pp. 1–11; Urteil 2H 168/44, pp. 1–8.

year 1944 and the end stages of the war, however, brought with them not just Neugebauer's missing 'common root of resistance' but along with it something new: a noticeable broadening of resistance away from its historical bases in the *KPÖ* and CCL entities and the emergence for the first time of significant numbers of both unorganised resisters and groups not tied to any one party or organisation. Some within this new wave, in fact, explicitly worked to escape the compartmentalisation of earlier resistance and actively built bridges across party lines. An example of such a group, New Free Austria (*Neues freies Österreich*), was formed in Freistadt in Upper Austria in late 1944. As its objective, it sought to bring individuals from both the left and right sides of the political spectrum together in a unified struggle to liberate Austria and end the war. To that end, the group began collecting and stockpiling weapons, food and money, lent support to allied advances into Austria and hoped to be in position to assist in the transfer of civil administration following the expulsion of the Nazis (Neugebauer 2013, pp. 266–67).

Other groups, such as the *Wiener Mischlingsliga*, established in 1943 at the encouragement of the *KPÖ*, had their origins instead in shared ethnicity. Led by Otto Andreasch, a *KJVD* functionary, its members were young, educated men and women, who, while all of at least partial Jewish ancestry, held a range of differing political views and stances. Before being rounded up by the *Gestapo*, the activities of the group extended to the production of pamphlets calling for armed insurrection against Nazi rule and at least 20 known acts of sabotage, among which included the damaging of aircraft engine parts and the delaying of trains full with munition factory employees on their way to work (Kuntz 2014, pp. 364–65).

Other late-war groups opted for more violent confrontations with the regime. An excellent example was the Austrian Freedom Front (*Österreichische Freiheitsfront* (*ÖFF*)), an anti-fascist organisation founded by Austrian and German communist refugees in occupied Brussels but purposely left open to fighters of all persuasions. In addition to participating in the Belgian Resistance, the organisation towards the end of 1943 established a company of Austrian-based partisans under the guidance of Erich Ungar. Mainly active in the Judenburg region of Upper Styria between late 1943 and early 1944, this group, numbering 23 at the time of its arrest for high treason, began a bombing campaign on war-related targets, such as railway junctions, and participated in a number of violent exchanges with soldiers and police, engagements often initiated with the sole aim of seizing weapons and munitions (Neugebauer 2013, pp. 253–54).

At the same time, another ongoing guerrilla campaign against German occupation was being waged by Slovenian partisans (Vogl 2009, pp. 159–75). With bases in the remote forests of the Slovenian highlands, their insurgency, which targeted German military and industrial assets, frequently spilled over into the Carinthian and Styrian borderland regions of Austria. There, they stockpiled weapons and food, created shelters and safe houses and operated networks of couriers. One such courier, Johann Klantschnik, was apprehended in May 1944 with 12 others and accused of carrying out an attack on a police station (Luza 1984, pp. 193–209; Vogl 2009, p. 173).[23]

[23] For an overview of this trial, see Baum (2011) and Baum (2012).

Table 5.7 provides an overview of the characteristics and background experiences of the most common types of resisters active in Austria. As in the case of Germany, given their weight of numbers, the communist sets the norm for what a typical Austrian resister looked like. In common with their German counterparts, the *KPÖ*, and leftist resisters in general, tended on average to be a primary-educated, blue-collar worker in their late-30s at the time of their arrest. One point of difference, however, was that they were much more likely to have had some prior history in a social democratic organisation than German communists. Indeed, the majority of Austrian communists – 64% of *KPÖ* and 73.7% of non-*KPÖ* communists – shared this background. It was a characteristic which had its roots in the *RSÖ*'s withdrawal from the resistance struggle in 1938 and the mass defection to the communists this engendered. As Neugebauer (2013, pp. 241–42) has previously observed, 'the overwhelming majority of resistance fighters who appeared before the court were former social democrats, who were active in communist organisations after 1938'. A further manifestation of this phenomenon is the relatively high proportion of left resisters of all shades who had served in the Republican Protection League (*Republikanischer Schutzbund*) in an unsuccessful attempt to defend the party and forestall the ever-deeper radicalisation of Austrian politics after World War I. One such resister was Wilhelm Jeindl, the married son of a turner with a basic primary school education; throughout the 1920s and 1930s, he held down a range of unskilled blue-collar jobs for short periods of time, in between more extended bouts of unemployment. He had four sons, three of whom by 1942 were serving on the Eastern Front. A member of the *SPÖ* from 1919, he joined the Republican Protection League in 1934, the Fatherland Front (*Vaterländische Front*) in 1937 and then, finally, in the summer of 1939 at the age of 36, the *KPÖ*. A member of a *KPÖ* cell active in Vienna between 1938 and 1941, he was eventually picked up by the *Gestapo* in April 1942 and sentenced to 10 years in a penitentiary for membership of an illegal organisation, separatism and the circulation of subversive materials (Anklage 7 J 285/42, pp. 1–5; 12–14; Urteil 5H 138/42, pp. 1–5).

Conservative resisters in Austria, likewise, had much in common with their German counterparts. Inter alia, they were also much better educated, more religious, more active in conservative and nationalists politics and more likely than other resisters to have served in World War II and be employed as either a professional, property owner, soldier or a state official. Yet, Austrian conservatives had their distinctive features as well. Age-wise, they were on average considerably younger than the German national conservative resister; indeed, in this regard, they found themselves fairly much in line with the rest of the Austrian resistance population. They were also more Roman Catholic in their religious beliefs, more likely to have partial Jewish ancestry and far less inclined to have been at any point of their lives in the Nazi Party, the *SA* or the *SS* (7.7% and 1.9% of them, respectively, in Austria compared to 14.4% and 7.2% in Germany). Finally, the focus of their political activity was – unsurprisingly – on peculiarly Austrian manifestations of conservativism and nationalism: the Christian Social Party (*Christlichsoziale Partei* or *CSP*), Legitimists organisations and Austrian nationalist organisations. Earlier in

Table 5.7 Personal experiences and characteristics of defendants arrested for activities in Austria, by resistance group (in percentages)

		KPÖ	Non-KPÖ communist	Other Left	CCL	Not organised	Overall
Education and training							
Highest educational attainment	Primary	95.3	99.2	91.7	48.1	86.8	88.1
	High school medium	1.4	0.8	1.7	3.8	0.0	1.7
	High school advanced	0.9	0.0	0.0	7.2	7.9	2.0
	Tertiary	2.3	0.0	6.7	40.9	5.3	8.2
Vocational training	Completed or in process	63.4	59.3	66.7	33.7	42.1	58.1
Citizenship and ethnicity							
Citizenship	German	3.0	0.8	13.3	0.0	2.6	3.6
	Austrian	93.9	98.3	85.0	96.2	89.5	93.4
	Foreign	2.7	0.0	1.7	2.4	2.6	2.3
	Stateless	0.4	0.8	0.0	1.4	5.3	0.7
Ethnicity	Partial Jewish ancestry	2.2	0.8	3.3	5.8	2.6	2.8
Occupation							
	Student	0.2	0.0	3.3	7.2	0.0	1.4
	No particular occupation	5.5	4.2	6.7	10.1	7.9	6.2
	Blue collar (skilled and unskilled)	58.4	65.3	56.7	11.5	73.7	52.0
	White collar (excluding high-ranked state officials)	6.7	4.2	8.3	16.8	5.3	8.2
	Professionals (excluding military)	1.8	0.0	3.3	26.0	2.6	5.3
	Business or property owner	0.8	0.0	0.0	4.8	0.0	1.3
	High-ranked state official	0.2	0.0	0.0	6.3	0.0	1.1
	Engaged in full-time political work	7.2	0.0	3.3	0.0	0.0	5.1
	Enlisted soldiers or non-commissioned military officer	1.2	0.0	0.0	4.8	0.0	1.6
	Commissioned military officer	0.0	0.0	0.0	0.0	0.0	0.0
	Service worker	17.9	26.3	18.3	12.5	10.5	18.0

Employment status							
	Employed	92.8	94.9	86.7	88.0	92.1	91.6
	Unemployed	0.4	0.0	1.7	1.4	0.0	0.8
	Not in the labour force	6.8	5.1	11.7	10.6	7.9	7.6
Age							
	Mean age at arrest	38	42	39	40	40	39
Experiences							
Personal and familial	Chronic health problem	7.3	7.6	10.0	13.5	23.7	9.0
	Spent time outside the Reich (in a private capacity)	8.2	4.2	15.0	13.9	5.3	9.3
	Engaged or married	73.7	78.8	65.0	43.3	60.5	68.7
	Widowed	1.7	0.8	5.0	3.8	2.6	2.1
	Divorced	3.5	3.4	5.0	3.4	2.6	3.5
	Has children	24.4	16.1	28.3	26.4	10.5	23.7
	Came of age during 1918–1919 Revolution	24.0	29.7	21.7	16.3	21.1	23.0
	Came of age during the hyperinflation	32.9	26.3	30.0	22.1	26.3	30.4
	Immediate family engaged in anti-regime activities	12.6	8.5	20.0	16.3	7.9	13.0
	Served in World War I	18.1	28.0	25.0	21.6	26.3	20.1
	Served in World War II	6.6	4.2	6.7	10.6	2.6	6.8
	Family member killed during war	3.2	3.4	1.7	5.3	2.6	3.4
Religious	Motivated by strong religious convictions	3.7	0.8	5.0	47.6	7.9	10.3
	Roman Catholic	3.6	0.8	5.0	47.1	7.9	10.2
	Evangelical	0.1	0.0	0.0	0.5	0.0	0.1

(continued)

Table 5.7 (continued)

		KPÖ	Non-KPÖ communist	Other Left	CCL	Not organised	Overall
Legal	Previous convictions	25.6	16.9	33.3	12.5	39.5	23.9
	Previous terms of Imprisonment	22.4	14.4	28.3	9.1	34.2	20.6
	Previous major political convictions	3.4	1.7	0.0	1.9	0.0	2.8
	Previous minor political convictions:						
	Defamation	2.7	1.7	5.0	1.9	2.6	2.6
	Public disorder	1.3	3.4	3.3	0.5	2.6	1.6
	Resisting state authorities	0.2	0.0	1.7	0.0	0.0	0.2
	Illegal meetings and/or writings	10.8	5.1	10.0	1.0	18.4	9.0
	Malicious gossip	0.4	0.8	1.7	1.4	0.0	0.6
	Previous non-political convictions:						
	Property crimes	5.2	4.2	8.3	1.9	13.2	5.0
	Sexual offences	0.3	0.0	3.3	0.5	2.6	0.5
	Crimes of violence	5.1	1.7	3.3	1.4	5.3	4.2
	Other crimes	5.0	1.7	8.3	5.8	7.9	5.3

Organisational memberships (current or past)	Denied entry/expelled from Hitler Youth/ League of German Girls	0.6	0.0	1.7	1.9	10.5	1.1
	Non-political and non-religious	14.3	19.5	18.3	6.3	5.3	13.4
	Trade union	41.2	54.2	41.7	8.2	39.5	37.1
	Free Corps/Association of Front Soldiers	2.1	2.5	3.3	11.5	0.0	3.6
	SA/SS	1.6	1.7	3.3	1.9	2.6	1.8
	Nazi Party (German or Austrian)	3.7	1.7	3.3	7.7	7.9	4.3
	Black Front	0.0	0.0	0.0	0.0	2.6	0.9
	Communist	47.5	12.7	1.7	1.0	36.8	34.9
	Social Democrat	64.0	73.7	56.7	7.7	44.7	54.9
	Trotskyist	0.0	5.9	0.0	0.0	0.0	0.5
	Anarchist	0.0	0.8	0.0	0.0	0.0	0.1
	Christian Social	0.9	0.0	0.0	11.5	2.6	2.5
	Protecting Austrian Republic	23.8	32.2	18.3	1.4	10.5	20.3
	Home Guard	1.8	5.9	0.0	14.9	0.0	4.0
	Fatherland Front	25.6	32.2	20.0	59.1	7.9	30.3
	Legitimist	0.0	0.0	10.0	49.5	0.0	7.7
	Austrian national conservative	0.0	0.0	0.0	29.3	0.0	4.3
	Pan-German Nationalist	0.9	0.0	0.0	2.4	2.6	1.1

Table 5.8 Regional pattern of activity by main resistance groups in Austria, 1938–1945 (in percentages)

	KPÖ/KPD	Non-*KPÖ*/KPD communist	Other left	CCL	Not organised
Burgenland	2.9	0.0	0.0	0.0	5.3
Carinthia	3.2	0.0	0.0	9.3	0.0
Lower Austria	10.5	8.5	0.0	3.9	2.6
Salzburg	5.5	0.0	34.1	6.9	2.6
Styria	20.7	66.9	8.5	14.6	60.5
Tyrol	3.6	0.0	0.0	8.4	2.6
Upper Austria	5.4	0.8	4.9	6.9	2.6
Vienna	47.3	23.7	35.4	47.2	23.7
Vorarlberg	0.9	0.0	17.1	3.0	0.0

their lives, many had served in the Home Guard (*Heimwehr*), too, while a clear majority (59.1%) had been members of Dolfuß's authoritarian Fatherland Front.

In terms of the location of resistance group activity in Austria, Table 5.8 reveals Vienna and Styria as vitally important foci for all the main resistance groups but especially the *KPÖ*, CCL and the unorganised resister. Table 5.8 also locates especially high concentrations of non-*KPÖ*/*KPD* communist and individual, non-organised activity in Styria and other left groups in Salzburg. As in the case of Germany, nonparametric statistical analysis of the provincial distribution of activity of the two largest groups of resisters in Austria, the *KPÖ* and CCL, indicates that the share of their resistance exhibited in any particular province did not reflect the proportion of their supporters nationally that resided in that province, as measured by the number of votes each had attracted there in the 1930 National Election.[24] However, while the share of the party's national vote in any particular province did not map directly onto the actual number of its resisters active there, the relative ranking of a province's share of the *KPÖ*'s national vote was a good – indeed, the best – predictor (with a rank correlation coefficient of 0.88) of a province's relative ranking in the levels of both overall and *KPÖ* resistance activity. Expressed in plain language, this finding suggests that those provinces which had larger numbers of *KPÖ* votes also tended to experience greater volumes of resistance activity. The intensity of resistance at the provincial level was, thus, broadly aligned with the number of *KPÖ* supporters who resided there.

[24] With the chi-squared test for conservative resistance, CCL activity is assumed to have been linked politically most closely to the vote of the *CSP*. Further details of the tests carried out are available from the authors on request.

5.3 Outside the Reich

As Nazi power spread, first in Germany itself and then subsequently outside, it set off waves of individuals fleeing its persecution and, in many cases, hoping to continue their resistance from safer climes. Most opted to seek refuge in countries that neighboured the Reich. Such proximity to Germany, it was hoped, would ensure that the exiled parties and groups could continue to exert influence there. But as time would soon painfully demonstrate, it was a flawed strategy that misjudged both the efficacy of the *Gestapo* in countering incursions from outside and, crucially, the speed with which Hitler would act on his expansionary ambitions. The series of rapid conquests and annexations the regime enjoyed meant that for many such exiles, their reprieve and relative security elsewhere would prove transitory. The journey of the exiled *SPD* leadership is illustrative of the plight experienced by many: between early 1933 and March 1937, it was based in Prague, from March 1937 to May 1940 in Paris and then from June 1940 in London (McDonough 2001).

Together, Tables 5.9 and 5.10 and Fig. 5.5 provide an overview of serious internal resistance to National Socialist rule outside Germany and Austria. The vast majority of outside resisters were affiliated with the *KPD/KPÖ* or the *SPD*, the only two groups which were able to remain active outside Germany and Austria for the entire duration of the Third Reich. While their arrests peaked in 1935, they together dominated resistance outside the Reich in every year other than 1938 and 1944. The *KPD/KPÖ* diaspora, like its *SPD* counterpart, spread across most parts of continental Europe but had heavy concentrations at different times in Czechoslovakia, the Netherlands, France and the USSR. Many of its cadre, including senior leaders such as Wilhelm Pieck and Walter Ulbricht, found refuge in Moscow.

Rudolf Hallmeyer provides a good example of the mobility experienced by resisters operating outside the Reich. A pipelayer by training, at the age of 16, he joined the *KJVD* in 1924 and then the *KPD* in 1931. Elected in December 1932 as a State Deputy for his Saxon hometown, Plauen, he went underground after the Reichstag fire. He escaped to Prague in April 1934 but returned soon after to undertake some work for the Central Committee in Hannover and Magdeburg. He was once again forced to flee Germany in the summer of 1935. On this occasion, he went to the Soviet Union and attended the Sixth World Congress of the Communist Youth International in Moscow and then embarked upon a 2-year stint at the International Lenin School. In 1938 and 1939, he undertook another series of assignments for the *KPD*, which involved him travelling between Prague, Gothenburg and Berlin. In 1940, he was sent to Copenhagen and then back to Berlin, where he established a cell in Berlin-Moabit. Not long after, though, the *Gestapo* finally caught up with him in August 1940 as he returned to Berlin from a series of meetings with contacts in Magdeburg, Halle and Dresden. Hallmeyer was convicted of high treason and guillotined in Plötzensee on 8 September 1943 (Anklage 10 J 17/41 g, pp. 1–6; Urteil 1H 194/43, pp. 1–3).

Table 5.9 Arrests for treason and high treason for activities outside the Reich, 1933–1945, by resistance group or organisation

	KPD/KPÖ	Other communist	*SPD*	Other socialist	Anarchist	Black Front	Buendische Youth	20 July	Not organised	Austrian CCL	Trade union
1933	12	7	12	0	1	0	0	0	0	0	0
1934	17	3	4	0	0	13	0	0	0	0	0
1935	102	0	42	2	0	26	9	0	0	0	0
1936	23	6	26	3	7	13	1	0	0	0	0
1937	19	11	0	4	0	2	1	0	0	0	0
1938	16	5	9	8	0	12	0	0	1	7	*0*
1939	49	0	4	1	0	8	8	0	2	15	*5*
1940	37	4	2	1	0	0	2	0	3	24	8
1941	76	2	3	1	1	0	0	0	4	8	1
1942	20	0	1	*6*	0	0	0	0	3	0	1
1943	34	0	4	0	0	0	0	0	5	0	0
1944	3	0	1	0	0	0	0	10	0	10	0
1945	0	0	0	0	0	0	0	0	0	0	0
Total	*408*	*38*	*108*	*26*	*9*	*74*	*21*	*10*	*18*	*64*	*15*

Table 5.10 Main areas of resistance activity outside the Reich, by main groups and organisations (in percentages by location of activity)

	KPD/KPÖ	Non-KPD/*KPÖ* communist	*SPD*	Other left	CCL and 20 July	Black Front
Belgium	3.9	7.5	21.3	14.0	8.4	0.0
Bulgaria	0.8	0.0	0.0	0.0	0.0	0.0
Czechoslovakia	29.7	42.5	36.0	32.6	23.4	82.3
Denmark	7.1	0.0	6.7	0.0	0.0	2.5
France	16.9	50.0	8.0	27.9	33.6	1.3
Greece	0.0	0.0	0.0	0.0	0.0	1.3
Hungary	0.5	0.0	0.0	0.0	34.6	1.3
Italy	0.2	0.0	0.0	0.0	0.0	0.0
Luxembourg	0.6	0.0	0.7	3.5	0.0	0.0
The Netherlands	18.5	0.0	26.0	11.6	0.0	10.1
Norway	1.8	0.0	0.0	2.3	0.0	0.0
Poland	1.8	0.0	0.7	4.7	0.0	0.0
Romania	1.0	0.0	0.0	0.0	0.0	0.0
USSR	15.9	0.0	0.7	3.5	0.0	0.0
Yugoslavia	1.4	0.0	0.0	0.0	0.0	1.3

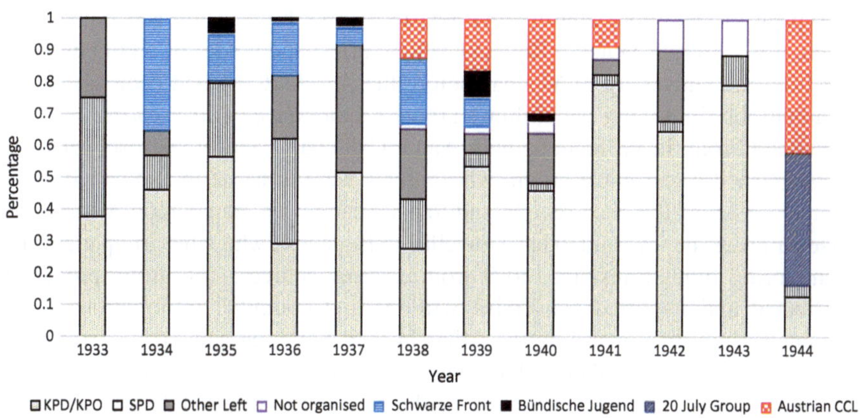

Fig. 5.5 Incidence of treason and high treason outside the Reich by main parties and groups, 1933–1945 (in percentages)

Not surprisingly, fewer *SPD* fleers ended up in the Soviet Union. Instead, Czechoslovakia, the Low Countries, France and Denmark were the sites of most of their external resistance.[25] For example, Franz Bott, a wirepuller from Cologne and *SPD* member since 1929, smuggled packages and party newspapers into and out of Germany for most of 1934. Operating mainly out of the Netherlands, it was also noted in his *Gestapo* files that he had spoken in August 1934 at a social democratic conference in Liege that had been critical of the Comintern's strategy of forming a unity front (*Einheitsfront*) of anti-fascist forces. Apprehended by the regime in May 1935, he was sentenced to 4 years in a penitentiary (Anklage 9 J 479/35, pp. 1–9; Urteil 2H 14/36, pp. 1–9; 58).

Other groups active and operational outside the Reich in the pre-war/early war period were the Black Front (1934–1939), especially in Czechoslovakia and the Netherlands, the Buendische Youth (1935–1940) and a range of smaller socialist and communist (1933–1942) parties like the *KPDO* and *SAP*. From approximately 1938 onwards, other groups began to make their presence felt, namely, unorganised resisters, trade unionists and, most significantly of all, Austrian Catholic-Conservative-Legitimists. The impact of this steadily growing diversification of groups willing and able to challenge the regime externally was that by 1944, left-wing resisters had for the first time ceased to make up the majority of arrests for treasonous or high treasonous activities outside the Reich.

Typical of this more conservative resister, active both inside and outside the Reich, was Heinrich Maier, a Roman Catholic priest and holder of a PhD in theology from the University of Vienna. Before the *Anschluß*, he had been associated with both the right-wing paramilitary group, *Ostmärkische Sturmscharen*, and the Fatherland Front. In 1938, along with Walter Caldonazzi and Franz Josef Messner, he established a resistance group, whose goals were to assist in and expedite the military defeat of the Nazi regime and to establish a free, democratic Austria. For these purposes, he was said to have travelled to Budapest, where his group may have been storing funds transferred to it from external sources. The group produced, distributed and imported subversive writings and compiled information on the armaments industry for use by the Allies in their bombing campaign. Arrested in March 1944 with nine others, Maier was executed on 22 March 1945. He was the last person to be executed in Vienna by the Nazis (Anklage 6 J 158/44, pp. 1–3; 18–19; Urteil 5H 96/44, pp. 1–3; 11–12).

Table 5.11 sheds light on some of the common attributes and characteristics of those engaged in resistance outside the Reich. As one might expect, they were, as a group, remarkably similar to comparable resisters active in Germany and Austria. Four differences are worth highlighting, however.

---

[25] It should be noted that the UK, neutral European countries and other parts of the world, including the USA, do not figure in Table 5.10. This is because exiled resisters in these locations were never apprehended by the Nazi state. These countries' absence for Table 5.10 should not, therefore, be taken to mean that no resistance occurred in those places. This was not the case. Indeed, the exiled leadership of the *SPD* based itself after the fall of France in London.

Table 5.11 Personal experiences and characteristics of defendants arrested for activities outside the Reich, by resistance group (in percentages)

		KPD/KPÖ	Non-*KPD/KPÖ* communist	*SPD*	Other Left	CCL and 20 July	Overall
Education and training							
Highest educational attainment	Primary	87.3	65.8	81.5	62.0	33.8	74.6
	High school medium	3.2	5.3	0.0	4.0	1.4	3.3
	High school advanced	3.2	5.3	5.6	8.0	9.5	6.2
	Tertiary	6.4	23.7	13.0	26.0	55.4	15.9
Vocational training	Completed or in process	60.0	55.3	44.4	34.0	21.6	48.4
Citizenship and ethnicity							
Citizenship	German	73.8	92.1	87.0	86.0	14.9	73.6
	Austrian	13.5	0.0	1.9	6.0	78.4	15.5
	Foreign	2.9	2.6	7.4	4.0	6.8	4.0
	Stateless	9.8	5.3	3.7	4.0	0.0	6.8
Ethnicity	Partial Jewish ancestry	2.9	7.9	3.7	8.0	8.1	4.0
Occupation							
	Student	1.0	0.0	0.9	4.0	10.8	2.9
	No particular occupation	4.2	0.0	6.5	6.0	5.4	4.4
	Blue collar (skilled and unskilled)	44.6	55.3	42.6	34.0	8.1	38.2
	White collar (excluding high-ranked state officials)	5.1	15.8	9.3	22.0	12.2	8.7
	Professionals (excluding military)	2.7	5.3	4.6	6.0	27.0	6.1
	Business or property owner	2.0	2.6	0.9	0.0	8.1	2.5
	High-ranked state official	0.0	0.0	0.0	2.0	2.7	0.4
	Engaged in full-time political work	26.7	2.6	9.3	2.0	0.0	15.5
	Enlisted soldiers or non-commissioned military officer	0.0	0.0	0.9	0.0	5.4	0.6
	Commissioned military officer	0.0	0.0	0.0	0.0	8.1	0.8
	Service worker	13.7	18.4	25.0	24.0	12.2	19.8

(continued)

Table 5.11 (continued)

		KPD/KPÖ	Non-*KPD/KPÖ* communist	*SPD*	Other Left	CCL and 20 July	Overall
Employment status							
	Employed	53.9	63.2	63.0	70.0	90.5	61.9
	Unemployed	13.7	26.3	20.4	2.0	1.4	12.5
	Not in the labour force	32.4	10.5	16.7	28.0	8.1	25.5
Age							
	Mean age at arrest	35	36	36	39	38	36
Experiences							
Personal and familial	Chronic health problem	8.1	5.3	6.5	6.0	17.6	8.8
	Spent time outside the Reich (in a private capacity)	21.8	21.1	17.6	30.0	23.0	23.3
	Engaged or married	51.2	39.5	65.7	54.0	50.0	53.1
	Widowed	1.2	2.6	1.9	6.0	1.4	1.6
	Divorced	9.6	13.2	2.8	8.0	4.1	7.7
	Has children	21.1	15.8	22.2	22.0	27.0	21.6
	Came of age during 1918–1919 Revolution	25.0	39.5	33.3	14.0	14.9	24.9
	Came of age during the hyperinflation	41.7	39.5	32.4	28.0	18.9	35.9
	Immediate family engaged in anti-regime activities	18.4	5.3	15.7	16.0	13.5	15.8
	Served in World War I	18.1	21.1	31.5	24.0	21.6	21.6
	Served in World War II	2.0	0.0	0.0	2.0	21.6	3.4
	Family member killed during war	1.5	0.0	0.9	0.0	5.4	1.4
Religious	Motivated by strong religious convictions	2.9	2.6	0.0	2.0	31.1	7.2
	Roman Catholic	1.7	2.6	0.0	0.0	29.7	6.1
	Evangelical	1.2	0.0	0.0	2.0	1.4	1.0

Legal	Previous convictions	40.2	21.1	24.1	32.0	10.8	32.9
	Previous terms of Imprisonment	35.0	15.8	17.6	26.0	4.1	27.4
	Previous major political convictions	10.8	5.3	1.9	6.0	1.4	7.1
	Previous minor political convictions:						
	Defamation	6.1	2.6	4.6	4.0	2.7	5.4
	Public disorder	5.9	0.0	0.9	4.0	0.0	3.8
	Resisting state authorities	2.7	0.0	1.9	4.0	0.0	1.9
	Illegal meetings and/or writings	8.1	5.3	1.9	2.0	0.0	4.9
	Malicious gossip	0.0	0.0	0.0	0.0	0.0	0.1
	Previous non-political convictions:						
	Property crimes	11.8	7.9	7.4	10.0	4.1	10.4
	Sexual offences	1.5	0.0	0.9	2.0	0.0	1.5
	Crimes of violence	7.4	0.0	4.6	10.0	0.0	6.3
	Other crimes	15.7	15.8	13.0	12.0	5.4	15.0
Organisational memberships (current or past)	Denied entry/expelled from Hitler Youth/League of German Girls	0.0	0.0	0.0	0.0	2.7	0.8
	Non-political and non-religious	17.4	15.8	24.1	16.0	4.1	15.8
	Trade union	30.6	34.2	42.6	56.0	8.1	29.2
	Hiking group	2.7	2.6	0.0	2.0	0.0	6.1
	Protecting Weimar Republic	1.2	2.6	0.0	12.0	0.0	2.9
	Free Corps/Association of Front Soldiers	1.7	0.0	4.6	6.0	10.8	5.4
	SA/SS	0.5	0.0	0.9	2.0	4.1	3.2
	Nazi Party (German or Austrian)	1.1	0	0	0	9.5	7.7
	German Communist	71.8	68.4	6.5	24.0	0.0	43.9
	German Social Democrat	17.4	28.9	78.7	62.0	0.0	26.0

(continued)

Table 5.11 (continued)

	KPD/KPÖ	Non-*KPD/KPÖ* communist	*SPD*	Other Left	CCL and 20 July	Overall
German Socialist	12.5	23.7	4.6	28.0	1.4	10.5
German national conservative	1.2	0.0	0.9	0.0	1.4	3.5
Centre Party	0.5	0.0	0.0	0.0	2.7	2.3
Anarchist	0.2	0.0	0.0	18.0	0.0	1.3
Trotskyist	0.0	26.3	0.0	0.0	0.0	1.3
Christian Social	0.2	0.0	0.0	0.0	4.1	0.6
Austrian Communist	7.8	0.0	0.0	0.0	1.4	4.2
Austrian Social Democrat	9.1	0.0	0.0	0.0	4.1	5.1
Protecting Austrian Republic	2.0	0.0	0.0	0.0	0.0	1.0
Home Guard	0.0	0.0	0.0	0.0	17.6	1.8
Fatherland Front	2.0	0.0	0.0	0.0	41.9	5.2
Legitimist	0.0	0.0	0.0	0.0	33.8	3.2
Austrian national conservative	0.0	0.0	0.0	0.0	20.3	1.9
Pan-German Nationalist	0.0	0.0	0.0	0.0	4.1	0.4

First, the resister operating outside the Reich tended to be younger than those inside it by a few years. This observation held for virtually all types of resisters but was most noticeable among the conservatively aligned, who were on average as much as 13 years younger than their counterparts active in Germany.

Second, resisters working outside the Reich were on the whole better educated. About 15.9% of them had had a tertiary education; the corresponding figures for Germany and Austria were 9.9% and 8.2%, respectively. Again, this characteristic held for all major groups of resisters other than the *KPD/KPÖ*.

Third, outsider resisters tended to be more ethnically diverse, with larger shares of people who were foreign, stateless or of partial Jewish ancestry. As much as 8.1% of conservative resisters, for example, were of mixed blood, more than two-and-a-half times the rate found among arrested conservative resisters in Germany.

Fourth, outside resistance on the left contained a significantly higher proportion of full-time party cadres than was the case among those arrested in both Germany and Austria. For example, 26.7% of *KPD/KPÖ* and 9.3% of *SPD* arrests for activities outside the Reich were party cadres. In Germany, the percentages were 13.4% and 5.0% and in Austria 7.2% and 3.3%. Among other things, these figures reflect the exiling of these parties' leadership following Hitler's ascension to power.

In concluding this section, it is important to emphasise that the exceptional challenges confronted by the outside resister should not be underestimated. Engaging in resistance in an alien setting away from one's home, family and culture and regularly crossing guarded frontiers no doubt added levels of complexity and danger to an act that in one's own country was already perilously risky. Together, the four differences identified by Table 5.11 suggest that those operating outside the Reich attempted to meet those challenges in part by being on average marginally more youthful, energetic, informed and deeply committed to their party and cause.

5.4 Conclusion

This chapter has outlined the contributions of the main resistance groups active in Germany, Austria and outside the Reich between 1933 and early 1945. While this exercise has highlighted the amazing diversity of political and spiritual beliefs which provided the organisational context for, and motivation to, those who chose to engage in serious resistance, it has also identified the enduring importance of certain parties and groups – in particular, the *KPD* and *SPD* in Germany and the *KPÖ* and Catholic-Conservative-Legitimist organisations in Austria – in that struggle to bring down Hitler's regime from within. In the next chapter, our focus shifts to what happened to serious resisters once they had been apprehended by the authorities and taken into custody.

References

Primary Sources

(A) Archival Files

(i) Widerstand als Hochverrat

Anklage 9J 479/35 In: *Nationalsozialismus, Holocaust, Widerstand und Exil 1933–1945*. Online-Datenbank. De Gruyter. Accessed May 23, 2013. http://db.saur.de/DGO/basicFullCitationView.jsf?documentId=wh2833

Anklage 17J 152/37 In: *Nationalsozialismus, Holocaust, Widerstand und Exil 1933–1945*. Online-Datenbank. De Gruyter. Accessed May 23, 2013. http://db.saur.de/DGO/basicFullCitationView.jsf?documentId=wh1065

Anklage 9J 355/39g In: *Nationalsozialismus, Holocaust, Widerstand und Exil 1933–1945*. Online-Datenbank. De Gruyter. Accessed May 8, 2013. http://db.saur.de/DGO/basicFullCitationView.jsf?documentId=wh2193

Anklage 8J 203/41 In: *Nationalsozialismus, Holocaust, Widerstand und Exil 1933–1945*. Online-Datenbank. De Gruyter. Accessed Feb 12, 2014. http://db.saur.de/DGO/basicFullCitationView.jsf?documentId=wh3022

Anklage 10J 17/41g In: *Nationalsozialismus, Holocaust, Widerstand und Exil 1933–1945*. Online-Datenbank. De Gruyter. Accessed Oct 11, 2013. http://db.saur.de/DGO/basicFullCitationView.jsf?documentId=wh3143

Anklage 7J 285/42 In: *Nationalsozialismus, Holocaust, Widerstand und Exil 1933–1945*. Online-Datenbank. De Gruyter. Accessed Feb 10, 2014. http://db.saur.de/DGO/basicFullCitationView.jsf?documentId=wh1530

Anklage 5J 120/43 In: *Nationalsozialismus, Holocaust, Widerstand und Exil 1933–1945*. Online-Datenbank. De Gruyter. Accessed Jan 29, 2014. http://db.saur.de/DGO/basicFullCitationView.jsf?documentId=wh1189

Anklage 6J 158/44g In: *Nationalsozialismus, Holocaust, Widerstand und Exil 1933–1945*. Online-Datenbank. De Gruyter. Accessed Feb 12, 2014. http://db.saur.de/DGO/basicFullCitationView.jsf?documentId=wh1971

Urteil 2H 14/36 In: *Nationalsozialismus, Holocaust, Widerstand und Exil 1933–1945*. Online-Datenbank. De Gruyter. Accessed May 23, 2013. http://db.saur.de/DGO/basicFullCitationView.jsf?documentId=wh2834

Urteil 2H 51/37 In: *Nationalsozialismus, Holocaust, Widerstand und Exil 1933–1945*. Online-Datenbank. De Gruyter. Accessed May 23, 2013. http://db.saur.de/DGO/basicFullCitationView.jsf?documentId=wh383

Urteil 2H 9/40 In: *Nationalsozialismus, Holocaust, Widerstand und Exil 1933–1945*. Online-Datenbank. De Gruyter. Accessed May 8, 2013. http://db.saur.de/DGO/basicFullCitationView.jsf?documentId=wh2194

Urteil 2H 89/40 In: *Nationalsozialismus, Holocaust, Widerstand und Exil 1933–1945*. Online-Datenbank. De Gruyter. Accessed Jan 29, 2014. http://db.saur.de/DGO/basicFullCitationView.jsf?documentId=wh2990

Urteil 5H 138/42 In: *Nationalsozialismus, Holocaust, Widerstand und Exil 1933–1945*. Online-Datenbank. De Gruyter. Accessed Feb 10, 2014. http://db.saur.de/DGO/basicFullCitationView.jsf?documentId=wh1464

Urteil 1H 194/43 In: *Nationalsozialismus, Holocaust, Widerstand und Exil 1933–1945*. Online-Datenbank. De Gruyter. Accessed Oct 11, 2013. http://db.saur.de/DGO/basicFullCitationView.jsf?documentId=wh3144

Urteil 2H 141/43 In: *Nationalsozialismus, Holocaust, Widerstand und Exil 1933–1945*. Online-Datenbank. De Gruyter. Accessed Jan 29, 2014. http://db.saur.de/DGO/basicFullCitationView.jsf?documentId=wh1190
Urteil 2H 168/44 In: *Nationalsozialismus, Holocaust, Widerstand und Exil 1933–1945*. Online-Datenbank. De Gruyter. Accessed Feb 12, 2014. http://db.saur.de/DGO/basicFullCitationView.jsf?documentId=wh3023
Urteil 5H 96/44 In: *Nationalsozialismus, Holocaust, Widerstand und Exil 1933–1945*. Online-Datenbank. De Gruyter. Accessed Feb 12, 2014. http://db.saur.de/DGO/basicFullCitationView.jsf?documentId=wh1972

(ii) Dokumentationsarchiv des österreichischen Widerstandes

Dokumentationsarchiv des österreichischen Widerstandes ed. 1975a. *Widerstand und Verfolgung in Wien 1934–1945: eine Dokumentation*, vol. 1. Vienna: Österreichischer Bundesverlag.
Dokumentationsarchiv des österreichischen Widerstandes ed. 1975b. *Widerstand und Verfolgung in Wien 1934–1945: eine Dokumentation*, vol. 2. Vienna: Österreichischer Bundesverlag.

(B) German Laws

Reichsministerium des Innern. 1933. Das Gesetz zur Wiederherstellung des Berufsbeamtentums. *Reichsgestzblatt* 7 April 1933, Part I, 175.

Secondary Sources

Baum, Wilhelm. 2011. *Die Freisler-Prozesse in Kärnten: Zeugnisse des Widerstandes gegen das NS-Regime in Österreich*. Klagenfurt: Kitab-Verlag.
Baum, Wilhelm. 2012. *Zum Tode verurteilt: NS-Justiz und Widerstand in Kärnten*. Klagenfurt: Kitab-Verlag.
Benz, Wolfgang. 2014. *Der deutsche Widerstand gegen Hitler*. Munich: C.H. Beck.
Botz, Gerhard. 1988. *Nationalsozialismus in Wien: Machtübernahme und Herrschaftssicherung*, 3rd edition. Buchloe: Obermayer.
Form, Wolfgang and Rolf Engelke. 1996. 'Hochverrat' – 'Heimtücke' – 'Wehrkraftzersetzung'. Zur politischen Strafjustiz in Hessen. In *Verfolgung und Widerstand in Hessen 1933–1945*, eds. Renate Knigge-Tesche and Axel Ulrich, 26–43. Franfurt am Main: Eichborn.
Gleißner, Ruth-Maria. 2004. *Der Hitler soll das Maul nicht so voll nehmen: das kurze Leben des Walter Klingenbeck*. Munich: Bayerischer Rundfunk.
Hagspiel, Hermann. 1995. *Die Ostmark: Österreich im Großdeutschen Reich 1938 bis 1945*. Vienna: Universitäts-Verlagsbuchhandlung.
Hartl, Rupert. 1995. *Wem gehört Hitler? Wohin gehört Österreich?* Steyr: Ennsthaler Verlag.
Holmes, Blair and Alan Keele. 1995. *When Truth Was Treason: German Youth against Hitler. The Story of the Helmuth Hübener Group based on the narrative of Karl-Heinz Schnibbe*, foreword by Klaus J. Hansen. Urbana: University of Illinois.
Holtmann, Karen. 2010. *Die Saefkow-Jacob-Bästein-Gruppe vor dem Volksgerichtshof. Die Hochverratsverfahren gegen die Frauen und Männer der Berliner Widerstandsorganisation 1944–1945*. Paderborn: Ferdinand Schöningh.
Hormayr, Gisela. 2012. "*Ich sterbe stolz und aufrecht*": *Tiroler SozialistInnen und KommunistInnen im Widerstand gegen Hitler*. Innsbruck: Studien Verlag.

Jagschitz, Gerhard. 1988. Von der 'Bewegung' zum Apparat: Zur Phänemenologie der NSDAP 1938 bis 1945. In *NS-Herrschaft in Österreich*, eds. Emmerich Talos, Ernst Hanisch and Wolfgang Neugebauer, 487–516. Vienna: Verlag für Gesellschaftskritik.

Jahnke, Karl Heinz and Alexander Rossaint. 2002. *Hauptangeklagter im Berliner Katholikenprozess 1937. Kaplan Dr. Joseph Cornelius Rossaint*. Frankfurt am Main: Verlag für Akademische Schriften.

Klause, Ekkehard. 2004. Konservative im Widerstand. In *Widerstand gegen die nationalsozialistischer Diktatur 1933*–1945, eds. Peter Steinbach and Johannes Tuchel, 185–201. Bonn: Bundeszentrale für Politische Bildung.

Konrad, Helmut. 1978. *Widerstand an Donau und Moldau. KPÖ u. KSC zur Zeit d. Hitler-Stalin-Paktes*. Vienna: Europaverlag.

Kuntz, Dieter. 2014. Jewish Resistance in Nazi Germany and Austria, 1933–1945. In *Jewish Resistance against the Nazis*, ed. Patrick Henry, 334–365. Washington D.C.: Catholic University of America Press.

Lange, Alexander. 2010. *Meuten-Broadway-Cliquen-Junge Garde: Leipziger Jugendgruppen im Dritten Reich*. Cologne: Böhlau Verlag.

Luza, Radomir. 1984. *The Resistance in Austria, 1938–1945*. Minneapolis: University of Minnesota Press.

Marxen, Klaus. 1994. *Das Volk und sein Gerichtshof: eine Studie zum Nationalsozialistische Volksgerichtshof*. Frankfurt am Main: Klostermann.

McDonough, Frank. 2001. *Opposition and Resistance in Nazi Germany*. Cambridge: Cambridge University Press.

Mehringer, Hartmut. 2004. Sozialdemokratischer and sozialistischer Widerstand. In *Widerstand gegen die nationalsozialistischer Diktatur 1933*–1945, eds. Peter Steinbach and Johannes Tuchel, 56–78. Bonn: Bundeszentrale für Politische Bildung.

Merson, Allan. 1985. *Communist Resistance in Nazi Germany*. London: Lawrence and Wishart Ltd.

Neugebauer, Wolfgang. 1988a. Das NS-Terrorsystem. In *NS-Herrschaft in Österreich*, eds. Emmerich Talos, Ernst Hanisch and Wolfgang Neugebauer, 163–184. Vienna: Verlag für Gesellschaftskritik.

Neugebauer, Wolfgang. 1988b. Widerstand und Opposition. In *NS-Herrschaft in Österreich*, eds. Emmerich Talos, Ernst Hanisch and Wolfgang Neugebauer, 537–552. Vienna: Verlag für Gesellschaftskritik.

Neugebauer, Wolfgang. 1989. *Österreicher und der Zweite Weltkrieg*, ed. Dokumentationsarchiv des österreichischen Widerstandes. Vienna: Österreichischer Bundesverlag.

Neugebauer, Wolfgang. 2013. Der österreichische Widerstand 1938–1945. In *Opferschicksale: Widerstand und Verfolgung im Nationalsozialismus*, ed. *Dokumentationsarchiv des österreichischen Widerstandes*, 233–72. Freistadt: Plöchl Druck GmbH.

Pelke, Else. 1961. *Der Lübecker Christenprozess 1943*. Mainz: Matthias-Grünewald-Verlag.

Peukert, Detlev, 1980. *Die KPD im Widerstand, Verfolgung und Untergrundarbeit an Rhein und Ruhr 1933–1945*. Wuppertal: Hammer.

Pohanka, Reinhard. 1997. *Pflichterfüller: Hitler's Helfer in der Ostmark*. Vienna: Picus Verlag.

Scholl, Inge. 1983. *The White Rose: Munich 1942–43*. Hannover: Weselyan Press.

Schlüter, Holger. 1995. *Die Urteilspraxis des nationalsozialistischen Volksgerichtshof*. Berlin: Duncker & Humblot.

Stadler, Karl. 1971. *Austria*. London: Ernest Benn Limited.

Steinbach, Peter and Johannes Tuchel eds. 1998. *Lexikon des Deutschen Widerstandes*. Munich: Verlag C.H. Beck.

Sturms, Frank. 2013. *Die Weiße Rose: Das Schicksal der Geschwister Scholl*. Wiesbaden: Marix Verlag.

Ueberschär, Gerd R. 2004. Der militärische Umsturzplan 'Walküre'. In *Widerstand gegen die nationalsozialistischer Diktatur 1933*–1945, eds. Peter Steinbach and Johannes Tuchel, 489–505. Bonn: Bundeszentrale für Politische Bildung.

Vogl, Josef. 2009. Ein Österreicher, der nur seine Pflicht getan hat. Markus Käfer und seine MitstreiterInnen im Kärntner Lavanttal. In *Schwerpunkt Bewaffneter Widerstand im Militär*, ed. *Dokumentationsarchiv des österreichischen Widerstandes*, 159–75. Vienna: LIT Verlag.

Wagner, Walter. 1974. *Der Volksgerichtshof im nationalsozialistischen Staat*. Stuttgart: Deutsche Verlags-Anstalt.

Walle, Heinrich. 2004. Der 20. Juli 1944. Eine Chronik der Ereignisse von Attentat und Umsturzversuch. In *Widerstand gegen die nationalsozialistischer Diktatur 1933*–1945, eds. Peter Steinbach and Johannes Tuchel, 505–521. Bonn: Bundeszentrale für Politische Bildung.

Weitz, Eric D. 1997. *Creating German Communism, 1890–1990: From Popular Protests to Socialist State*. Princeton: Princeton University Press.

Zarusky, Jürgen and Hartmut Mehringer eds. 1998. *Widerstand als Hochverrat 1933–45. Die Verfahren gegen deutsche Reichsangehörige vor dem Reichsgericht, dem Volksgericht und dem Reichskriegsgericht*. Munich: K G Saur. Microfiche edition.

Chapter 6
Crimes and Punishments

Over the course of the last three chapters, many aspects of the backgrounds, beliefs and actions of those who bravely resisted Nazism between 1933 and 1945 have been described, analysed and discussed. In this chapter, we turn our attention to another feature of their story, one that has typically been less scrutinised and certainly is not as well understood: the experience of the resister once they had been arrested and entered custody. After all, it was there in the interrogation chambers, cells and court rooms of the Nazi regime that the personal consequences of their actions were to be first painfully brought to bear.

The chapter begins with an outline of the nature and structure of the Nazi legal system and how it evolved to deal with the most serious of political crimes it confronted: treason and high treason. It then moves on to explore aspects of the reality of the resister's actual engagement with that system. In particular, the chapter considers the length of time it took to get from arrest to sentence, the type of charges resisters were likely to face, the verdicts they could expect, the determinants of the sentence they were ultimately to receive and what role, if any, the discretion of the judge presiding over the case would play in that process.

The chapter also investigates a critical choice that all resisters in custody, irrespective of their background or motivations, had to make, a choice that would directly impact their and their co-defendants' fate: should they cooperate with the authorities or not? Given that all who were arrested for serious resistance both fervently opposed the Nazi regime and were placed under enormous duress during interrogation, duress which normally included torture, why did some choose to cooperate, when others did not? Was it simply a matter of character or did other factors related to the resister's personal background, age, beliefs and affiliations influence their choice as well? Statistical analysis of the choices actually made by resisters in this situation offers rare insight on this question and indicates who among them were more or less likely to hold fast in the face of such horrendous pressure and pressing self-interest.

W. Geerling, G. Magee, *Quantifying Resistance*, Studies in Economic History,
DOI 10.1007/978-981-10-6008-3_6

6.1 Crimes

In all societies, high treason and treason rank as the most serious of political offences. High treason relates to those acts that internally undermine the power and integrity of the state, such as an attempt against the life of the head of state. Treason, by contrast, concerns itself with acts which harm the state externally, as in the rendering of assistance and provision of state secrets to a foreign power. When the Nazis came to power in January 1933 and began to use the courts to prosecute their political opponents, Articles 80–92 of the Federal Penal Code (*Reichsstrafgesetzbuch*), first promulgated in 1871, defined the acts of high treason and treason in Germany. According to the Code, there were three categories of high treason: (i) attacks that threatened the head of state, (ii) attempts to change or alter the territory of the Federal state and (iii) violations of the constitution. The Nazis, however, considered these laws as antiquated, excessively liberal and too narrow.[1]

On the night of 27 February 1933, Germany's parliament, the *Reichstag*, mysteriously went up in flames. The fire was attributed by the Nazis to the illegal work of communists and portrayed as a sign of an imminent armed uprising. It also supplied the rationale for the enactment of two decrees the following day which extended the scope of political crimes and sanctions for high treason in Germany. The first of these was the Decree of the President against Treason and High Treasonous Activities (*Die Verordnung des Reichspräsidenten gegen Verrat am Deutschen Volke und hochverräterische Umtriebe*) of 28 February 1933, which aimed to neutralise political opponents of the state, especially the Communist Party of Germany (*Kommunistische Partei Deutschlands* or *KPD*). The concept of subversive high treason (*Zersetzungshochverrat*) was introduced and defined as applying to all 'acts of political subversion, which obstructed the police and army from fulfilling their duty to protect the state' (Reichsministerium des Innern, *Reichsgesetzblatt* (*RGBl*) 28 February 1933b, Part I, p. 85). Under Article 6, the manufacture, distribution and storing of subversive writings were also, for the first time, included as acts of high treason. Article 1 of the other decree promulgated on the same degree, the Decree of the President for the Protection of People and State (*Die Verordnung des Reichspräsidenten zum Schutz von Volk und Staat*), commonly known as the *Reichstag Fire Decree*, extended the possibility of capital punishment to all high treasonous offences, including those which had previously carried a maximum penalty of life imprisonment (Reichsministerium des Innern, *RGBl* 28 February 1933a, Part I, p. 83). The complete political ascendancy of the Nazi Party was further tightened 5 months later when the Law against the Reestablishment of Parties (*Das Gesetz gegen die Neubildung von Parteien*) recognised the Nazi Party as the sole legitimate political party in Germany. Thereafter, the establishment, or reestablishment, of any political party became a high treasonous offence, punishable by a

[1] For good overviews of criminal (and political) justice in Nazi Germany more generally, see Müller (1987), Evans (1996) and Wachsmann (2004).

prison sentence of up to 10 years (Reichsministerium des Innern, *RGBl* 14 July 1933c, Part I, p. 479).

The *Reichstag* fire and subsequent trial also provided the impetus for the restructuring of the German judicial system. When the Nazis came to power, cases of high treason and treason were tried before the Supreme Court (*Reichsgericht*), a court which had held this responsibility since 1879. Accordingly, the *Reichstag* fire trial was conducted before the Supreme Court from September through December 1933. On 23 December 1933, its verdict was handed down. While the main defendant, Marinus van der Lubbe, was sentenced to death, the cases against the trial's other four defendants were dismissed due to lack of evidence (Wagner 1974, pp. 14–15).

The acquittal of four defendants in a trial that had attracted worldwide attention was an embarrassment to the regime, particularly given that this crime had been used as proof of a grand communist conspiracy and the pretext for the *Reichstag Fire Decree*. Hitler later described the verdict as 'laughable' (*lächerlich*) and the judges as 'senile' (*vertrottelt*) (Picker 1951, pp. 241–43). On the day on which the verdict was handed down, the Nazi Party's official newspaper *People's Observer* (*Völkischer Beobachter*) launched a scathing criticism of the verdict and proposed an alternative approach to interpreting the law more in touch with popular sentiment:

> This judgment is a judicial error, which perhaps more than anything else shows the necessity of a fundamental reform to our system of law...If the verdict had been delivered in accordance with genuine law, that … has its roots in healthy sentiment, then the sentence would have been pronounced differently. In that case, of course, the entire basis and conduct of the trial, which the people followed with growing displeasure, would have been altogether different. (Cited in Wagner 1974, p. 868)

The Ministry of Justice had in fact resisted calls immediately following the fire to introduce a court to pass judgement on matters in connection with the Reichstag fire on the grounds that Article 105 of the constitution made this unconstitutional. Only a constitutional amendment could make such a course of action possible (Gruchmann 1988, pp. 958–59). The growing volume of disparaging remarks from the press and leading party officials, however, pushed the Ministry to take more urgent action. On 29 January 1934, Hitler's State Secretary Hans Heinrich Lammers wrote to Dr. Franz Gürtner, Minister of Justice, criticising the mild punishments handed out for cases of treason and suggesting that such cases be tried before Special Courts (Wieland 1989, p. 19). In a cabinet meeting of 23 March 1934, Adolf Hitler, Ernst Röhm, Hermann Göring, Franz Gürtner, Wilhelm Frick, Minister of Defence Werner von Blomberg and State Secretary Hans Pfundtner discussed the establishment of a special People's Court to pass judgement on cases of high treason and treason. It was agreed at this time that the court would consist of two professional judges and three lay judges (Wagner 1974, p. 959). Gürtner was instructed to draw up the decree.

On 24 April 1934, that decree, the Law Amending Provisions of Criminal Law and Criminal Procedure (*Das Gesetz zur Änderung von Vorschriften des Strafrechts und des Strafverfahrens*), formally created the People's Court and transferred jurisdiction of high treason and treason to it from the Supreme Court. The law regulated the jurisdiction of the People's Court, the composition of its senates, the appointment of members and certain procedural details (Reichsministerium des Innern, *RGBl* 24 April 1934, Part I, p. 345).[2] Like its predecessor in trying cases of high treason and treason, each of the three senates of the People's Court consisted of five members for main trials (three for other trials), where each member of the senate, along with the presiding judge, decided on the case before them. Unlike the Supreme Court, though, only the presiding judge and one assessor in any major trial before the People's Court were required to be professional judges. The other three judges were honorary and drawn from the Nazi Party and its ancillary organisations. These non-jurists were intended to provide 'political expertise' and were praised in the founding law as being 'closely attached with the politically dynamic strength of the people whose popular conception of legality they best reflect' (Reichsministerium des Innern, *RGBl* 24 April 1934, Part I, p. 345). While it was hoped that each trial would be conducted on a collegiate basis, with each member required to sign the final decision, in practice the court operated according to the *Führer* principle, with the presiding judge dominant (Koch 1989, p. 53). The role of other members of the panel in the trial was purely advisory; they neither voted on guilt nor decided on the punishment imposed. These were the prerogative of the presiding judge, who also provided the formal reasoning for his decision. As Roland Freisler, soon to become President of the Court, explained in 1941:

> Next to the prosecution it would be the [presiding] judge's task to see to it that everything at issue would be cleared up swiftly and thoroughly. He was to lead and to decide; fellow professional as well as lay judges representing the *Volk* could advise but not decide. Responsibility lay with the presiding judge alone. (Cited in Koch 1989, p. 81)

Members of the People's Court, including its President, were appointed directly by Hitler for terms of 5 years (Gruchmann 1988, pp. 960–61).[3] According to Wilhelm

[2] Although jurisdiction over cases of high treason was transferred to the People's Court in 1934, minor cases continued to be sent to the Higher Regional Court for trial. In the period of Nazi rule before the People's Court was functional (January 1933 to July 1934), the Supreme Court heard cases against 314 defendants charged with treason and high treason. 129 of these defendants appear in our sample. Our number is lower because we have excluded cases which had some of their key files missing or illegible or which began prior to January 1933. On the structure of the People's Court, see Zarusky and Mehringer (1998, pp. 30–31). Good legal histories of the People's Court can be found in Lauf (1994), Marxen (1994), Schlüter (1995), Richter (2001), Marxen and Schlüter (2004), Wagner (1974) and Köpke (2011).

[3] It is worth noting that Hitler's role in appointing judges to the People's Court contrasted with the situation with respect to the Supreme Court, where appointments were made by the President of the Reich, Paul von Hindenburg, till his death on 2 August 1934. The original personnel appointed to the People's Court included 12 professional judges and 21 honorary assessors or lay judges. Of the legal members, three were Senate Presidents (*Senatspräsidenten*), including Fritz Rehn, the first President of the People's Court, six presiding Judges from a division of a District Court (*Landesgerichtsdirektoren*), one Judge of a District Court (*Landesgerichtsrat*), one Judge of a Local Court (*Amtsgerichtsrat*) and one Director of Public Prosecution (*Oberstaatsanwalt*). On 18 September 1934, shortly after the People's Court's establishment, Fritz Rehn died of a gall bladder disease. The position of President of the Court remained vacant until the appointment of Otto Thierack at the beginning of May 1936. In the interim, Wilhelm Bruner acted as its managing Vice President.

Weiß (1935, p. 518), lay member of the People's Court and later Chief Editor of the *Völkischer Beobachter*, the purpose of the new court was to form an institution which 'portrayed an essential part of the National Socialist jurisdiction, an organic creation of the National Socialist state, since the assessors formed a close personal relationship with the Party which did not exist in any other German court'.

As for those who appeared before the People's Court, until the last months of the war, each was required by law to be represented by a defence counsel. Under Section 3 of Article 4 of the April 1934 decree, however, the choice of defence counsel had to be approved by the presiding judge. That permission could be withdrawn at any time, too (Marxen 1994, p. 32). Theoretically, a defendant was to be represented either by their own selected defence lawyer (*Wahlverteidiger*) or a court-assigned counsel (*Pflichtverteidiger*). Given that the defendant's choice was conditional on the court's approval and politically unreliable lawyers had already been removed from the bar, the principle of free advocacy stipulated in the law was largely an illusion. Hans Richter (1934, p. 606), Ministerial Adviser in the Ministry of Justice, conceded as much in an article published in *Deutsche Justiz* on the day when the People's Court was first enacted: 'In cases where the state entrusts the highest court on account of their importance for the general public, care should be taken that only defence counsels who enjoy general trust be involved'. In any case, a raft of court regulations, blocking or severely restricting inter alia the defence counsel's access to the indictment, court files, the defendant, the evidence and the witnesses, ensured that no real defence could be mounted. Furthermore, a lawyer who made an unfavourable impression on the court or took the interest of their client too seriously could expect to be removed from the case and threatened with the withdrawal of their licence to practice (Wagner 1974, pp. 33–37). As a consequence, defence counsel had no or little bearing on proceedings. The quality of the defence counsel provided was never a factor in the verdict or sentencing of any of the defendants who appeared before the People's Court.

The law that established the People's Court also regulated the series of codes and decrees passed in the period since 1933 that had Nazified the Federal Penal Codes. Article 83 of the Federal Penal Code focused on key aspects of high treason. The death penalty was permitted for the following crimes: (i) the establishment, or reestablishment, of an organisation conspiring to commit high treason; (ii) the subversion of police and defence forces; (iii) the influencing of the masses through writings, records, pictorial images and radio broadcasts; and (iv) the attempt to import high treasonous writings, records or pictorial images. Incitement of others to commit high treason or the conspiracy to commit high treason carried a maximum imprisonment term of 10 years, though the added proviso that 'under certain conditions a lifelong prison sentence or the death penalty could be imposed' made the distinction between the crimes irrelevant (Reichsministerium des Innern, *RGBl* 24 April 1934, Part I, pp. 341–348). In short, in the People's Court, the judges had absolute discretion in the sentence imposed for treason and high treason up to, and including, the death sentence.

With the adoption of the provisions of April 1934, a solitary all-embracing law now brought all decrees with respect to high treason and treason into line. The death penalty was applicable to all crimes listed in Articles 80–83. According to the founding law, the People's Court decided in the first and last instance in all matters previously under the jurisdiction of the Supreme Court. No appeal against a decision was possible. Furthermore, the presiding judge was given significant latitude in determining punishment, as the sentencing options were elastic, both with respect to the type and severity of punishment handed down. Indeed, although Judges' Letters (*Richterbriefe*), first circulated by the Ministry of Justice in October 1942, reminded them to take into account the healthy sentiment of the people, there were no specific guidelines on sentencing (Boberach 1975). Judges were in effect free to operate off a sentencing scale that increased in severity from a term in a normal prison (*Gefängnis*) at one end, through time with hard labour and the loss of civil rights in a penitentiary (*Zuchthaus*), to capital punishment at the other. Lengths of imprisonment were also at the discretion of the presiding judge (Reichsministerium des Innern, *Reichsgesetzblatt* 24 April 1934, Part I, pp. 341–348).

On 1 August 1934, the first trials took place before the three senates of the People's Court. In April 1936, the Law about the People's Court and the Twenty-Fifth Amendment to the Salary Law (*Das Gesetz über den Volksgerichtshof und über die fünfundzwanzigste Änderung des Besoldungsgesetzes*) raised the status of the court to that of 'an ordinary constitutional court, standing alongside the Supreme Court', and a senior Nazi Party figure, Otto Georg Thierack, was named as its new President (Reichsministerium des Innern, *RGBl* 20 April 1936a, Teil I, p. 369). Holding that position until he left to become the Reich Minister for Justice in August 1942, Thierack was instrumental in shaping the fledgling court's procedures and practices to ensure that they accorded to the will of its founders. More important than the personnel changes was the symbolic decision to give lifetime tenure to the President of the Court, who also presided over the first senate, the three other Senate Presidents and six full-time judges (Reichsministerium des Innern, *RGBl* 20 April 1936a, Teil I, p. 369). The People's Court was now to be the living, and seemingly permanent, embodiment of Nazi justice.

During the Thierack presidency (1 May 1936–19 August 1942), the power of the People's Court continued to grow. Three primary reasons lay behind its growth. First, the annexation of Austria and the Sudetenland extended its territorial reach. Second, new crimes were steadily added to the jurisdiction of the court—all of which carried the penalty of death. The failure to report an intended crime, for example, insofar as that crime was high treasonous or treasonous, fell within the purview of the People's Court following the amendment to Article 139 of the Federal Penal Code in 1936. Likewise, the Law against Economic Sabotage (*Die Verordnung gegen Wirtschaftssabotage*) extended the charge of high treason to 'anyone who deliberately and unscrupulously, for personal gain or other low motives, illegally smuggled property abroad or left property abroad, thereby inflicting damage on the German economy' (Reichsministerium des Innern, *RGBl* 1 December 1936c, Part 1, p. 999). Subversion or undermining the fighting spirit of the German armed forces through the promotion and encouragement of desertion, inducement to disobedience or disobedience also became a capital offence with the promulgation of the Special Wartime Penal Code (*Die*

Kriegssonderstrafrechtsverordnung) on 17 August 1938 (Reichsministerium des Innern, *RGBl* 17 August 1938, Part 1, pp. 1455–57). Third, in response to these new responsibilities and the increased workload they engendered, the size of the court itself expanded and grew more complicated in its organisation. At its inception, the court had three senates: Senates 1 and 2 were responsible for high treason cases and Senate 3 for treason cases. As the number of cases steadily increased, a fourth senate—specialising in treason cases—was established from November 1935. On 1 November 1941, a fifth senate was added and in December 1942, a sixth senate, both of which heard cases of treason and high treason (Wagner 1974, p. 26; Koch 1989, p. 46). In practice, however, this apparent division of labour between senates proved largely illusory. Over its history, Presidents of the Court were to redefine the responsibilities of each of the senates on numerous occasions. From the annual business plans of the court, one can identify between 1934 and 1945 at least two major changes of foci for each senate, with the exception of Senate 6, the last to be created. Thus, the allocation of duties between senates for 1939 shows Senate 1, originally a pure high treason senate, taking on cases of treason for the first time. All high treason and treason cases from Austria and cases of economic sabotage were also added to its jurisdiction at this time. In addition, Senate 2 was now assigned cases based on Section 5 of the first decree from 28 February 1933 (arson, use of poison gas, floods and damage to railways) and Senate 4 with cases involving damage to military property. From October 1942, Senates 3 and 4 became pure treason senates; from 1 January 1943, only Senate 3 was confined to cases of treason. A short time later, however, Senate 3 had to take on a wider variety of cases, too. Moreover, by the latter stages of the war, Senates 5 and 6 had taken over responsibility for most cases from Austria and Bavaria (Wagner 1974, p. 46).

The jurisdictional confusion created by these shifting foci of the senates was later compounded by the regular intervention of Freisler in his role as President of the Court. During his presidency, the highest profile cases from the regime's perspective were assigned to Senate 1 (his senate) or Senate 2, irrespective of their nature and circumstances. He also reserved for himself the right to have the last word in all cases in which there were any discrepancies or uncertainty as to jurisdiction. In effect, such interventions served only to override and obfuscate the precise responsibilities of the different senates.[4] As Wagner (1974, p. 26) noted in this respect, 'the multiplication of senates, the constant growth in the number of cases, the fact that particular cases involved both high treason and treason, as well as the Freisler type of leadership brought with it an ever stronger blurring of jurisdiction within the individual senates'.

The onset of war brought about an even further expansion in the jurisdiction of the People's Court. The extension of the German Reich into Eastern Europe and the associated removal of German and Austrian opposition there ensured a steady stream of high treason cases, particularly from the annexed territories. New legislation extended the definition of high treason in two directions: specific war legisla-

[4] Freisler's actions eventually induced the Minister of Justice, Thierack, in November 1943 to demand an end to the arbitrary jurisdictional extension of Senate 1 (see Wagner (1974, pp. 26–27)).

tion in the form of the Decree Concerning Extraordinary Radio Measures of 1 September 1939 (*Die Verordnung über außerordentliche Rundfunkmaßnahmen*) and the Decree against National Parasites of 5 September 1939 (*Die Verordnung gegen Volksschädlinge*), both passed in the first week of the war with the aim of securing the home front from pernicious enemy propaganda, internal unrest or any act which could undermine the German war effort. Listening to foreign broadcasts and/or sending reports of foreign news, actions which in times of peace were not considered high treasonous, were strictly forbidden and subject to the death penalty (Reichsministerium des Innern, *RGBl* 1 September 1939a, Part 1, p. 1683; *RGBl* 5 September 1939b, Part 1, p. 2378). The National Parasites Decree defined three new capital crimes: (i) looting in 'liberated' territories, (ii) crimes during air raids and (iii) crimes of public danger. Furthermore, any other criminal act deemed by authorities to have 'exploited the extraordinary conditions caused by war' was punishable by death 'if the popular sentiment of the people requires it because of the particular wickedness of the act' (Reichsministerium des Innern, *RGBl* 5 September 1939b, Part 1, p. 2378).

Another piece of legislation, the Decree Concerning the Jurisdiction of Criminal Courts, Special Courts and other Criminal Procedures of 21 February 1940 (*Die Verordnung über die Zuständigkeit der Strafgerichte, die Sondergerichte und sonstige strafverfahrensrechtliche Vorschriften*), permitted a defendant who had committed more than one offence subject to different jurisdictions to be tried exclusively before the People's Court (Reichsministerium des Innern, *RGBl* 21 February 1940, Part 1, p. 405). Given that subversion of the armed forces, espionage and sabotage came under the jurisdiction of the People's Court (or were soon about to) as either cases of treason or high treason, this decree was extremely significant in establishing the primacy of the People's Court. The Decree Concerning the Expanded Jurisdiction of the People's Court of 10 December 1941 (*Die Verordnung über die erweiterte Zuständigkeit des Volksgerichtshofes*) provided greater access to cases of espionage, previously under the jurisdiction of the Supreme Military Court (*Reichskriegsgericht*); likewise, in March 1942, economic sabotage of armament industries through untrue reports of German economic strength or supplies came under the attention of the People's Court (Reichsministerium des Innern, *RGBl* 10 December 1941, Part 1, p. 776; *RGBl* 21 March 1942, Part 1, p. 165).

Two other legal reforms passed within the first 6 months of the war further curtailed the rights of the accused. Until this point in time, the People's Court—like other Special Courts—decided in the first and last instance. A defendant who escaped the death penalty could not be brought back before the same court for resentencing. The Law Amending Provisions of General Criminal Procedure, Defence Force Criminal Procedure and the Penal Code of 16 September 1939 (*Das Gesetz zur Änderung von Vorschriften des allgemeinen Strafverfahrens, des Wehrmachtsstrafverfahrens und des Strafgesetzbuches*) replaced this process through the creation of a special senate to deal with 'extraordinary appeals' from the Chief Public Prosecutor. Within a year of a sentence being handed down, the Director of Public Prosecution at the People's Court or Supreme Court could launch an appeal on the grounds that the verdict was wrong or the sentence was too lenient

(Reichsministerium des Innern, *RGBl* 16 September 1939c, Part 1, p. 1841). Like previous Nazi initiatives such as protective detention, 'extraordinary appeal' was a cynical term which permitted the arbitrary misuse of power, allowing the Ministry of Justice to enforce the political will of the regime. The special senate was characterised as the 'Senate of the Führer' and consisted of the President or Vice President of the Court, a professional associate judge at the court and three lay judges. According to Günther Wieland (1989, pp. 80–82), a special hearing always resulted in a draconian sentence.[5]

As Germany's fortunes in the war deteriorated, the People's Court, like the Nazi system as a whole, underwent an increasing radicalisation. Total war required that the court dispense expedient political justice against internal and external enemies of the state. As Joseph Goebbels' speech of 22 July 1942 made clear, 'Judges must base their decisions less on law and more on the basic idea that the law-breaker be removed from the national community. In war it is not so important whether a judgment is just or unjust, but whether it is appropriate' (cited in Broszat 1959, p. 438). Thierack thanked Goebbels for these sentiments and assured him that in future the judiciary would help implement these fundamental principles (Wieland 1989, p. 76). Less than a month later, Thierack would depart the People's Court to assume the position of Minister of Justice. He was replaced by Roland Freisler, a man who would give substance to the words of Goebbels.

An ideological zealot, Freisler, worked steadfastly as President to apply the National Socialist concept of law to the People's Court. As a result, more than anyone else, he has come to embody in the popular mind the court's wanton criminality, injustice and terror (Ortner 2013; Steinweis and Rachlin 2013). And as Nazi law became more radical and demanded harsher measures, this self-described political solider was only too happy to oblige and expunge internal and external enemies of the people. The concepts of high traitors, traitors and enemies of the people were eventually stretched to include all opponents and critics of the National Socialist regime. In his period as President (20 August 1942 to 3 February 1945), the death penalty became more common (Wagner 1974). On 3 February 1945, the People's Court building on Bellevuestrasse was badly damaged in a USAAF (United States Army Air Force) bombing raid in the course of which Freisler was killed (Buchheit 1968, p. 275). With the Red Army closing in on Berlin, a new President, Harry Haffner, was hastily appointed on 12 March 1945.[6] Under increasingly difficult circumstances, four more trials were held, and three more death sentences pronounced before the People's Court finally ceased its activities on 24 April 1945, the 11th anniversary of the court's founding. On 20 October 1945, by proclamation of the Allied Control Council in Germany, the People's Court was formally dissolved (Wagner 1974, p. 29).

---

[5] In 1944 alone, 90% of appeals resulted in the imposition of a death sentence (see Gruchmann (1988, pp. 1077–1081)).

[6] Between 4 February and 11 March 1945, Wilhelm Crohne acted as the managing Vice President of the court.

6.2 Punishments

For perfectly understandable reasons, the focus of most resistance studies has overwhelmingly fallen on the actions of key individuals or groups leading up to and during acts of resistance. The questions typically posed about resisters do not extend far beyond who they were, what they did and what they hoped to achieve through their actions. While these are crucial questions that we have touched on ourselves in previous chapters, it is important to remember that the story of resistance rarely ends with the act of resistance itself. Rather, for most resisters, the act, or attempted act, signalled the beginning of a different, often more protracted, and more personally challenging stage of the experience that included their arrest, interrogation, indictment, trial and punishment. These features of the resister's experience have been less systematically studied.

Available court records, however, do allow some light to be shed. For one thing, they reveal that this stage for most was a lengthy one. For the average defendant charged with treason and high treason, it took 431 days, or just over 14 months, from the day of their arrest to receive their sentence. There were, of course, exceptions, but, as Fig. 6.1 shows, it is interesting how stable this standard processing time remained across the entire pre-war period. In 1939 and 1940, though, the length between arrest and sentencing rose sharply to over 2 years (670 and 753 days, respectively), presumably as external resisters, brought once again back within the reach of the Nazi authorities by military successes, began to strain the resources of the legal system. The legal dimension of the process—viz. the time between indictment and sentencing—in fact, had been on the rise since 1937, growing from 120 days in that year to 346 days by 1939. Yet by 1941, the gap between indictment and sentencing had once again fallen back to its pre-war levels and continued to decline slowly thereafter, a decline that in part must have been due to the introduction of new senates of the People's Court in November 1941 and December 1942. By contrast, the period between arrest and indictment—the investigative and interrogative phase of the process—remained at around 11 months throughout the pre-war period, rising on average to as much as 562 days in 1940 and then falling precipitously from 1941. By 1944, the average time between arrest and indictment had declined to just 95 days, in other words, about 3 months. The reason for this steep decline in the time taken to get to an indictment is not clear. While it may have become easier for the *Gestapo* over time to detect and build cases against defendants because of the benefits of accumulated intelligence, greater resources being made available or just its willingness to use torture more readily, we have seen no direct evidence that would allow us to confirm this.[7] It is also possible that as the

[7] *Gestapo* numbers grew from around 1000 employees in 1933, to 6500 in 1937 and 15,000 by the outbreak of war. They continued to increase during the war and only reached their peak at the end of 1944 when 32,000 were employed by the *Gestapo*. As for torture, according to McDonough (2015, pp. 10, 43), 'during the latter stages of the war, the *Gestapo* became much more brutal in the way it treated the "enemies of the state" and "enhanced interrogation techniques" were used far more extensively'. It is possible that this greater use of torture in the final years of the war contributed to the reduction in the time needed to build a case against a defendant.

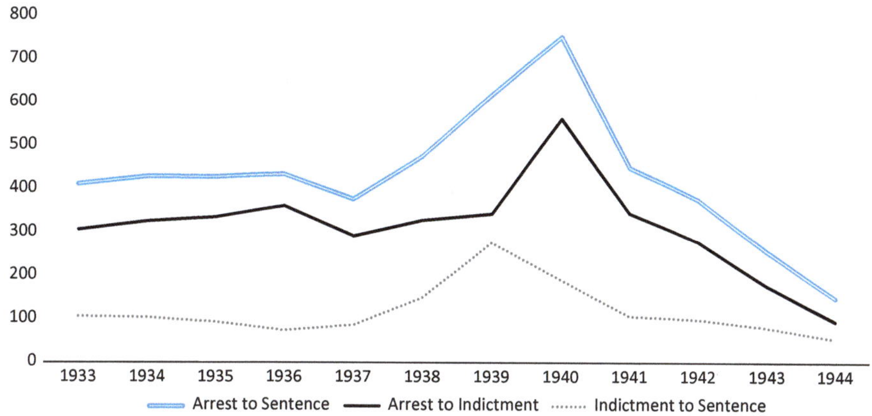

Fig. 6.1 Mean processing times of defendants (1933–1944) (Note: Mean times were calculated from a sample of 10% of each year's defendants)

radicalisation of the Nazi legal system deepened, especially from 1942, the required standards for cases to proceed to indictment and trial diminished. Whatever the reason, the unfolding wartime trends in the investigative and legal processing times complemented each other and ensured that the overall mean processing time of defendants had tumbled to just around 5 months by 1944. Of the two forces at play, Fig. 6.2 indicates that developments and practice in the investigative phase played the larger role in determining the length of the gap between arrest and sentencing, accounting for at least two-thirds of the processing time in each year other than 1939 when there was an approximate parity between the two phases.

Figure 6.3 presents a time series of the number of verdicts per month handed down to defendants in cases of treason and high treason. It shows, as did the annual series presented earlier in Fig. 3.1, that the number of proceedings and verdicts rose in the last few years of the war from fairly low volumes (on average about 19 verdicts per month between January 1936 and December 1941) to a peak in the latter half of 1944 of 112 verdicts per month. The monthly data, however, reveal a further dimension to the story that the annual series partially obscures, namely, that this upward trend in the volume of proceedings, which culminated in 130 sentences being handed down in November 1944, had actually already begun by mid-1942. So much so, in fact, that in August and September 1942—well before the defeat at Stalingrad—as many as 154 verdicts were reached by the People's Court, a volume that made these 2 months the most active months in the court's history until that point.

As Fig. 6.4 indicates, the vast majority of these verdicts—just under 92%—were convictions. Eight per cent of defendants, however, were either acquitted or had the case against them dismissed. The rate of conviction did not alter that significantly in

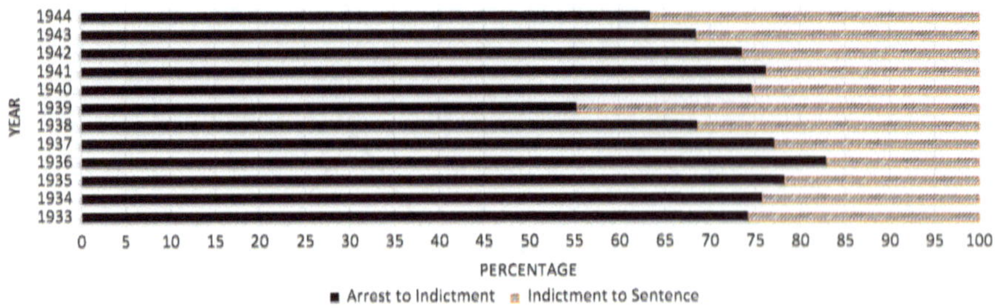

Fig. 6.2 Decomposition of mean processing times (1933–1944) (Note: Mean times were calculated from a sample of 10% of each year's defendants)

wartime; it was approximately 90% in the pre-war period, rising to just 93% during the war.

Of course, obtaining an acquittal or having a case discontinued or suspended in Nazi Germany did not necessarily mean release. In fact, it was commonplace for such individuals to be taken into 'protective custody' following discharge (Wachsmann 2004, p. 171).[8] Nonetheless, this rate of non-convictions in the People's Court may seem to be somewhat higher than expected. Unfortunately, cases which were discontinued, suspended or which produced acquittal verdicts tended not to generate the same trial records as those which had led to conviction. Qualitative analysis of the cases available, however, reveals that the most common reason for acquittals appears to have been the absence of evidence connecting a defendant to the resistance activities of their co-defendants. Erich Berger, a construction engineer and veteran of the Battle of France, provides a good example of such a case. He was arrested in July 1941 with five others on charges of belonging to an illegal organisation (the *KPD*), distributing illegal writings and engaging in subversive activities. The main defendant in the case, Kurt Gittel, was a communist with previous convictions for both high treason (in October 1934) and property damage. Following his release in 1936, Gittel resumed his activities with an illegal *KPD* group in Berlin. Based on evidence acquired from Gittel's interrogation, it was alleged by the *Gestapo* that Gittel had recruited Berger to his group and that Berger had given him a book about Lenin to read and discuss. Berger, however, denied all allegations: he had never belonged to a political party, had no knowledge of the group's subversive activities, had never given Gittel a book about Lenin and had never had discussions about communism with either Gittel or any other of his co-defendants. When the case went to trial, Gittel retracted his earlier incriminating testimony about Berger. As no other direct evidence of Berger's involvement with the *KPD* or Gittel's group could be adduced, Berger was acquitted (Anklage 6 J 166/41 g, pp. 1–4; Urteil 1H 137/41, pp. 1–4; 12–13).

[8] In the Reichstag fire trial, for example, the main defendant, Marinus van der Lubbe, was sentenced to death. His four co-accused were acquitted and then taken into protective custody (see Gruchmann (1988, p. 958)).

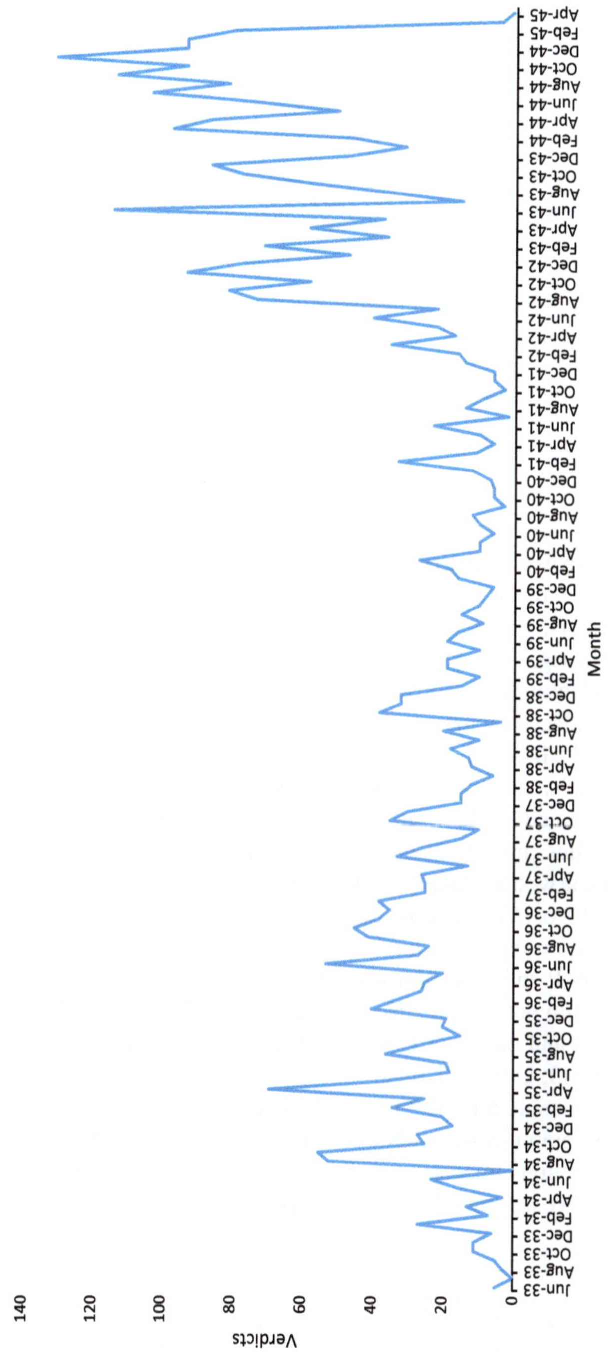

Fig. 6.3 Treason and high treason verdicts by month (January 1933–April 1945)

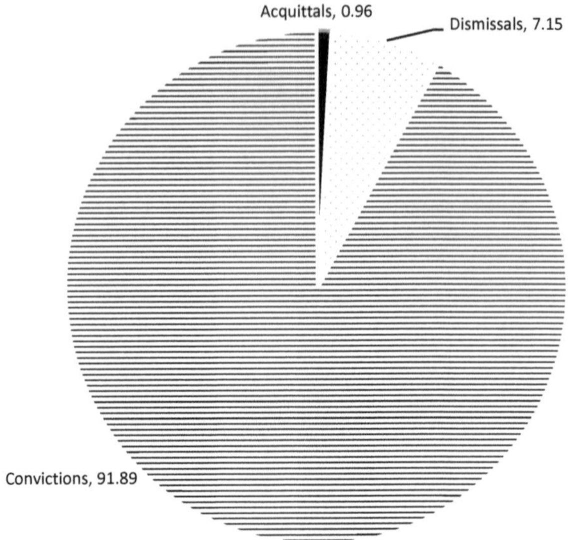

Fig. 6.4 Outcome of trials (1933–1945) (in percentages)

The arrest of wives and girlfriends of male resisters on the erroneous assumption that they too must have been engaged in resistance activities was another common source of acquittals. Charlotte Uhrig, for example, the wife of the communist resister, Robert Uhrig, was arrested in September 1942 on suspicion of involvement in the European Union (*Europäische Union*). When no evidence substantiating the suspicion was produced by the prosecution, she was acquitted (Anklage 11 J 32/44, pp. 1–7; Urteil 2H 51/44, pp. 1–2).

As for dismissals, cases were most typically discontinued or suspended when a defendant was deemed by the court to be afflicted by a serious physical or psychological illness. Thus, high treason charges against Josef P., a 30-year-old casual worker from Vienna who worked for the Communist Party of Austria (*Kommunistische Partei Österreichs* or *KPÖ*), were dismissed in 1943 on grounds of his mental impairment.[9] A court-appointed medical expert testified to the fact that at the time of his crimes Josef P. suffered from schizophrenia and congenital feeblemindedness and, therefore, could not have realised the illegality of his actions. The case against him was dropped, and on the orders of the court, he was confined to a sanatorium (Anklage 7 J 305/43; Urteil 5H 1/44).

Dismissals also could arise from official amnesties on certain minor crimes. Karl Böhm, a former communist from Niedereinsiedel in the Sudetenland, was arrested and charged with high treason after being caught with firearms on his person. Forensic testing revealed that the weapons had not been fired. When his case came

[9] The *Datenschutzgesetz* (protection of privacy law) places restrictions on the publishing of private information in Germany and Austria. In this book, we use a psychiatric evaluation of the defendant prepared by a psychiatrist commissioned by the People's Court. If the individual concerned is still living or has died within the last 30 years, permission is needed to reference personal material. We are not sure whether the individual concerned is still alive; therefore, all references to him in this book will be Josef P.

to the People's Court in the second half of 1939, it ruled that as Böhm had lost his father in the First World War, had no prior convictions, had made a full confession, had abandoned his dalliance with communism and had been charged with a crime not punishable by more than 2 years of imprisonment, he was covered by an amnesty of 7 June 1939. The case against him was dismissed (Anklage 5 J 686/38, pp. 1–4; Urteil 1H 59/39, pp. 1–3; 13–15).

These acquittals and abandoned cases notwithstanding, the vast majority of trials did lead to convictions. Table 6.1 breaks down these convictions by specific charges. Within the high treason category, it distinguishes between major charges like an attempt to assassinate the head of state and minor (in other words, less serious) charges such as the failure to report to the authorities others who may have engaged, or have planned to engage, in high treasonous activities. Table 6.1 reveals that over three quarters of convictions were for acts of major high treason, the most important of which were membership of an illegal organisation and involvement in the writing, production or distribution of anti-regime pamphlets, newspapers or flyers. These two varieties of high treason were responsible for more than half of convictions. Not surprisingly, Table 6.1 confirms that resistance activities by those smuggling materials into and out of Germany or by exiled opponents of the regime were at their most important (between 14 and 17% of convictions) in the pre-war and early war periods. From the end of 1940, however, the territorial expansion and occupations of the Third Reich effectively eradicated this form of activity. By the last year of the war, it accounted for just 1% of convictions. The charge of subversion referred to a broad category of activities which sought to interfere with the police and armed forces carrying out their duties. Throughout the duration of the regime, this typically sat at between 4 and 5% of convictions until 1943 at which point it more than doubled, after jurisdiction for undermining the fighting spirit of the German armed forces (*Wehrkraftzersetzung*) was transferred from the Supreme Military War Court to the People's Court. Convictions for separatism, almost exclusively handed down to those opposing the annexation of Austria, grew markedly over the early stages of the war, pinnacling in 1941 and 1942 and then dropping off steadily in the last 2 years of the regime. Listening to foreign broadcasts, which was not illegal prior to the war, constituted 3.5% of convictions, whereas charges of involvement in acts of violence against Adolf Hitler was an even rarer occasion, with just four convictions before the 20 July plot of 1944.

Around 10% of all convictions were on charges for acts of comparatively minor high treason. Interestingly, it was a variety of charge more common at the beginning and end of the regime than in its middle years. This pattern may reflect the fact that charges such as conspiracy to commit high treason (far more commonly used in the early years of the regime), aiding and abetting and failure to report (the two minor charges preferred in the last year of the war) were effective legal tools to round up and place in custody broad-based networks of opposition, conspiracies, as well as all who associated with them without necessarily having any hard evidence of their actual resistance activity. However, once the support base and organisation of the larger oppositional groups had been deracinated, the need to employ such charges became less pressing.

Table 6.1 Convictions for treason and high treason by charge (absolute number and share of all convictions) (1933–1945)

	Pre-war		Early war		Barbarossa to Stalingrad		Stalingrad to D-Day		Post-D-Day		Wartime (1939–1945)		Overall (1933–1945)	
	No.	%	No.	%	No.	%	No.	%	No.	%	No.	%	No.	%
Major high treason	**2316**	***83.3***	**570**	***88.4***	**1473**	***87.9***	**1566**	***75.6***	**1175**	***63.0***	**4784**	***76.4***	**7100**	***78.6***
Illegal organisation	1038	*37.4*	215	*33.3*	547	*32.6*	714	*34.5*	509	*27.3*	1985	*31.7*	3023	*33.4*
Illegal writings	763	*27.5*	158	*24.5*	410	*24.5*	342	*16.5*	210	*11.3*	1120	*17.9*	1883	*20.8*
Activities abroad	406	*14.6*	110	*17.1*	119	*7.1*	44	*2.1*	19	*1.0*	292	*4.7*	698	*7.7*
Subversion	108	*3.9*	31	*4.8*	70	*4.2*	191	*9.2*	210	*11.3*	502	*8.0*	610	*6.7*
Separatism	1	*0.0*	56	*8.7*	275	*16.4*	186	*9.0*	80	*4.3*	597	*9.5*	598	*6.6*
Violence against Hitler	0	*0.0*	0	*0.0*	1	*0.1*	3	*0.1*	63	*3.4*	67	*1.1*	67	*0.7*
Foreign broadcasts	0	*0.0*	0	*0.0*	51	*3.0*	86	*4.2*	84	*4.5*	221	*3.5*	221	*2.4*
Minor high treason	**393**	***14.1***	**29**	***4.5***	**71**	***4.2***	**179**	***8.6***	**263**	***14.1***	**542**	***8.7***	**935**	***10.3***
Conspiracy	313	*11.3*	9	*1.4*	18	*1.1*	38	*1.8*	16	*0.9*	81	*1.3*	394	*4.4*
Aiding and abetting	73	*2.6*	13	*2.0*	40	*2.4*	76	*3.7*	123	*6.6*	252	*4.0*	325	*3.6*
Failure to report	7	*0.3*	7	*1.1*	13	*0.8*	65	*3.1*	124	*6.6*	209	*3.3*	216	*2.4*
Treason	**70**	***2.5***	**46**	***7.1***	**132**	***7.9***	**327**	***15.8***	**428**	***22.9***	**933**	***14.9***	**1003**	***11.1***

Finally, the charge of treason—the betrayal of one's country to another—grew steadily from approximately one in 40 convictions before the war to nearly one in four from June 1944. While initially this charge tended to be used for Germans and Austrians caught working for, or supplying classified information to, foreign intelligence agencies—activities which naturally increased as the war approached and progressed—another important reason for the increasing prevalence of treason convictions from mid-1942 was the decision of the regime following the invasion of the Soviet Union to treat all communist resisters and fellow travellers as foreign operatives (Wagner 1974, p. 804). The charge was also increasingly applied as the war progressed to anyone of any persuasion who sought any form of unauthorised contact, planned or actual, with officials or citizens of a foreign state.

As Table 6.2 shows, treason was the charge that, upon conviction, brought the harshest penalties. 78.5% of those convicted of it were executed, and those who were not received prison or a penitentiary sentence noticeably longer than those convicted of high treason. Major high treason was also harshly dealt with: roughly, one-third of those convicted of it received a death sentence, and over 50% were sent to serve a lengthy term—on average over 11 years in duration—doing hard labour in a penitentiary. A comparatively shorter term in a normal state prison (usually around a year and a half), the most lenient of punishments handed out by the People's Court, was an outcome more common among individuals convicted of minor high treason. By contrast, only 1.7% of those found guilty of a minor high treason were sent to their death.[10]

Table 6.2 also illustrates a dimension of the growing radicalisation of the Nazi legal system. Prison and penitentiary terms steadily rose across the pre-war period and reached their zenith at the end of 1942. The mean prison term for someone convicted of treason, for example, grew from approximately two and a half years in the pre-war period to 12 years between June 1941 and the end of 1942. From 1943, prison and penitentiary sentences tended to become less frequent and shorter in duration. This late-war trend, however, was not due to the greater leniency of the court. Rather, it came about because serious convictions that had previously led to prison or penitentiary terms were increasing resulting in death sentences. Only less severe or incompletely established cases now avoided execution. As a consequence, the average terms of incarceration actually started to fall. By 1944, those sent to prison or a penitentiary were receiving sentences similar in length to those convicted before the war.

Figure 6.5 traces more precisely the mounting use of capital punishment by the People's Court. In the pre-war period, the death sentence was relatively uncommon: on average only 0.26 executions per month were ordered by the court or, in other words, about three a year. Between September 1939 and April 1942, the rate rose to 1.3 death sentences per month. The use of execution, however, rapidly expanded from May 1942 after the last vestiges of legal impartiality remaining, even in theory,

[10] During the pre-war period, 29.5% of convicted defendants were sent to prison, 69.1% were sent to a penitentiary and 1.4% received the death penalty. The corresponding figures for wartime were 9.0%, 46.0% and 45%, respectively.

Table 6.2 Sentences by charge, type of resistance and period (in percentages) (1933–1945)

	Pre-war	Early war	Barbarossa to Stalingrad	Stalingrad to D-Day	Post-D-Day	Wartime (1939–1945)	Overall (1933–1945)
Major high treason							
Prison	18.6	7.2	2.2	2.8	5.2	3.8	8.8
Penitentiary	79.6	82.0	37.0	45.7	42.9	46.2	57.5
Death	1.8	10.8	60.8	51.4	51.9	50.0	33.7
Mean prison term (months)	21.0	21.8	40.7	48.4	29.3	34.6	24.9
Mean penitentiary term (months)	81.6	105.3	133.0	85.1	70.6	95.0	88.7
Treason							
Prison	10.0	0.0	1.5	0.0	1.6	1.0	1.6
Penitentiary	80.0	67.4	6.8	6.1	19.6	15.4	19.9
Death	10.0	32.6	91.7	93.9	78.7	83.6	78.5
Mean prison term (months)	30.4	0.0	84.0	0.0	36.0	46.7	39.6
Mean penitentiary term (months)	126.1	140.6	188.0	134.4	75.6	104.8	110.8
Minor high treason							
Prison	66.7	51.7	17.1	32.6	29.8	30.2	45.3
Penitentiary	33.3	48.3	77.1	67.4	65.6	66.8	53.0
Death	0.0	0.0	5.7	0.0	4.6	3.0	1.7
Mean prison term (months)	18.8	15.3	23.5	21.9	22.2	21.5	19.9
Mean penitentiary term (months)	44.6	62.6	89.7	66.1	63.7	68.3	62.2
Violent resistance							
Prison	24.8	0.0	0.0	12.1	6.3	6.3	13.8
Penitentiary	74.3	55.6	28.6	3.0	17.9	20.0	42.0
Death	0.9	44.4	71.4	84.8	75.8	73.8	44.2
Mean prison term (months)	20.5	0.0	0.0	78.0	30.0	49.2	28.3
Mean penitentiary term (months)	58.9	180.0	99.0	144.0	63.5	106.9	72.5
Non-violent resistance							
Prison	26.7	12.1	3.5	7.9	12.0	8.5	15.4
Penitentiary	61.2	72.9	37.5	43.5	41.4	44.3	50.7
Death	1.3	7.3	56.0	41.1	36.1	40.0	25.3
Mean prison term (months)	19.9	18.7	34.7	27.6	22.8	25.0	21.7
Mean penitentiary term (months)	79.1	100.4	128.0	81.5	69.0	90.7	85.4

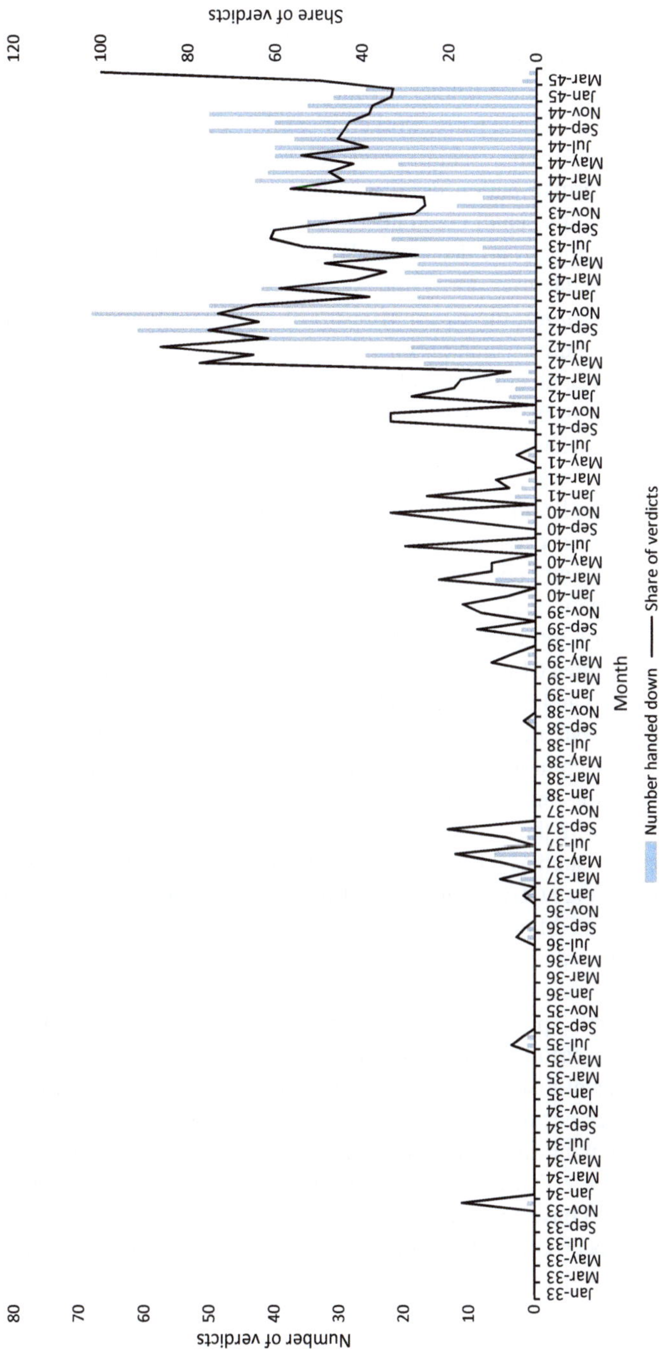

Fig. 6.5 Death sentences by month (number handed down and as a share of verdicts) (January 1933–April 1945)

were removed with the decision of the Reichstag to grant unrestricted powers to Hitler on 26 April 1942. As holder of supreme judicial power, Hitler demanded and received the power to 'force with all means at his disposal every German, if necessary, whether he be common soldier or officer, low or high official or judge, leading or subordinate official of the party, worker or employee, to fulfil his duties' (Cited in Wagner 1974, p. 804). This right overrode all existing legal regulations. Under the pretext that the German nation faced a struggle for its very existence, Hitler and those who acted on his behalf could now fire judges and hold to account those who had not reached appropriate decisions (Wagner 1974, p. 804; Wachsmann 2004, pp. 214–15). Desperate to restore Hitler's confidence in the judiciary and willing to work towards the *Führer*, the Reich Ministry of Justice responded through a series of radical new measures which promised swift, retributive justice. In May 1942, it determined that since all communist high treason possessed the character of treason, it should, henceforth, be normally punishable as such, that is, with death (Wagner 1974, p. 804). Accordingly, between May and December 1942, the number of death sentences handed down rose sharply to over 40 a month (about 70.9% of all sentences issued by the People's Court)—a rate (and share of overall sentences) higher than that experienced even at the height of the sentencing of the 20 July plotters between July 1944 and February 1945, when on average 39 individuals received the death sentence each month (39.3% of all sentences in that period). In terms of sheer volume, the use of the death sentence peaked in the months of September and December 1942, when 61 and 68 defendants, respectively, received that punishment. The unprecedented and unrepeated scale of the use of the death sentence in the latter half of 1942 suggests that the granting to Hitler of unconditional powers and the reclassification of communists as treasonous traitors in May 1942 may not have been the only factors influencing sentence patterns in that period. Another possible factor, consistent with the statistical analysis of sentencing discussed below, was that unexpected events such as the military setbacks in Russia earlier in the year and the heavy Royal Air Force (RAF) bombing of Luebeck, Hamburg, Rostock and Cologne between March and May 1942 may have engendered such a sense of shock and crisis in the regime that, in order to restore balance, it felt a need to retaliate and handle those who resisted its dominance, both externally and at home, more ruthlessly.[11] The notable spikes of the latter halves of 1942 and 1944 aside, the death sentence for high treason and treason in the final years of the war did settle at a much higher baseline level than ever before. From May 1942 to April 1945, the issuance of a death sentence occurred on average 30.4 times a month.

Even after allowing for the increasing harshness of sentences handed down, it is clear from Table 6.2 that at all times the sentencing decisions of People's Court judges appeared to have reflected the gravity of the charges for which a defendant before them had been convicted. Thus, those convicted of aiding and abetting could expect to be treated more leniently than those convicted of treason. Moreover,

[11] Wachsmann (2004, p. 210) writes 'In the following months [start of 1942], the public mood was punctured further by increased rationing and Allied air raids, with the first major British mass bombing raid on civilian targets hitting Lübeck on 28–29 March 1942'.

People's Court judges also seemingly distinguished between those who had engaged in non-violent and violent forms of resistance. The latter were more likely to be executed or receive longer prison or penitentiary terms.

How can we account for the exercise of such judicial discretion in a highly politicised court? After all, the People's Court, before which defendants charged with treason and high treason appeared, is widely regarded, as the post-war 'Justice Case' at Nuremberg labelled it, 'a terror court, notorious for the severity of punishment, secrecy of proceedings, and denial to the accused of all semblance of judicial process' (United States Government Printing Office 1951, p. 19). A closer reading of the regime's intentions for the court suggests an answer. From the time of the 'seizure of power' (*Machtergreifung*), the Nazi regime sought to use the courts as means to facilitate and expedite the prosecution and elimination of its political opponents. To that end, the Nazi leadership, ever keen to appropriate for itself the legitimacy of Germany's inherited legal traditions and confident of its ability to control the central institutions of society, was willing to give its judges significant latitude to act. And it was a trust that, from the regime's perspective, was seemingly not misplaced. Econometric analysis of sentencing patterns confirms that on the whole People's Court judges did perform in a manner consistent with Nazi expectations.[12]

The existence of this degree of judicial freedom, however, did inject a distinct element of heterogeneity into the sentencing process. As Table 6.3 shows, not all sources of political opposition received equal treatment from the Nazi court system. Rather, the courts operated in a more nuanced manner, targeting more ruthlessly those resisters whom judges believed posed the greatest challenges.

From the regime's perspective, the greatest and most imminent threat clearly lays on the left, in particular the two mass left-wing parties of the period, the *KPD* and the Social Democratic Party of Germany (*Sozialdemokratische Partei Deutschlands* or SPD). In the early years of the regime, these were the political movements best positioned to challenge Nazi supremacy.[13] Later on, however, the state's focus fell overwhelmingly on the *KPD*. The Nazi regime, in fact, saw itself in a life and death struggle, both inside and outside the Reich, with Bolshevism and all its manifestations. Having such connections on one's record brought severe retribution for defendants, markedly so after the invasion of the Soviet Union. In this regard, the written judgements of People's Court judges are revealing. While the same clichés about the war and the national community used against all defendants appear, the language used in judgements against communists was distinctly harsher. After May 1942, their verdicts were all prefaced with the declaration: 'the defendant is forever dishonourable'. Nor did communists have a place in the new order: 'They must also be kicked out of the national community, from which they renounced their place a

---

[12] This section on sentencing draws on work conducted by the authors elsewhere. For fuller information on, and discussion of, the econometric techniques used and the estimations on which our conclusions are based, see the following: Geerling et al. (2013, pp. 209–234), Geerling et al. (2016, pp. 517–542) and Geerling et al. (2017a, b). (Available at SSRN: https://ssrn.com/abstract=)

[13] For an overview of the major political parties in the 1930s, see Weitz (2009, pp. 81–127).

long time ago. Therefore, they are deprived of their civil rights'. Loss of civil rights was on par with the arbitrary use of judicial power. 'Such behaviour could only be punished with death even without legal regulation', read the same verdict. In sentencing four communists to death in a different trial, the court invoked the 'protection of the people' clause, but added 'That [the death penalty] is not only prescribed in the law, but corresponds to the healthy sentiment of the people and its need for protection from the inner enemy'. This need came ahead of protecting a defendant who was only 17 years old when arrested (Anklage 7 J 300/43, pp. 1–2; Urteil 5H 94/43, pp. 1–2; 12–13; Anklage 7 J 302/43, pp. 1–2; Urteil 5H 104/43, p. 51).

Those holding a leadership role in a left-wing party attracted the harshest punishment from judges. From this array of left-wing leaders of different leanings, *KPD* cadres, who numerically constituted the vast majority, emerged as those most ferociously targeted by the courts and, as a result, were among the most likely to receive death sentences or longer terms in penitentiaries. Around 47% of *KPD* and Communist Party of Austria (*Kommunistische Partei Österreich* or *KPÖ*) cadres, for example, were executed (the overall execution rate for defendants in our sample was just over 26.5%).[14] We also find that *KPD* and *KPÖ* leaders were less likely to be treated softly and sent to prison (4.5% were sentenced to prison terms compared to 15.3% of all defendants). Rank and file members of the *KPD* and *KPÖ* did not fare well either. While, statistically speaking, not so openly singled out for harsh treatment as their leaders, normal members of the *KPD* and *KPÖ* nonetheless still faced a disproportionate chance of capital punishment. As Table 6.3 shows, more than half of them were to be sentenced to death. They were also more likely to receive longer terms in a penitentiary than anyone else.

By contrast, affiliates of smaller left-wing organisations, Catholic groups, right-wing parties and national-conservative groups in both Germany and Austria, which were perceived by the authorities to pose a lesser threat to the state, were typically less likely to be executed and more likely to receive the most lenient of sentence types: a term in a normal state prison. Thus, between July 1941 and April 1945, 20.1% of Austrian Catholic-Conservative-Legitimists and 11.1% of German national conservatives were executed at a time when the overall rate of execution for those convicted for treason and high treason was 45.5%. At the same time, 17.7% and 11.1% of the members of Austrian and German conservative groups, respectively, received prison terms; the national rate was 8.0%. Defendants with records of involvement in right-wing militias and associations like the Free Corps (*Freikorps*) and Association of Front Soldiers (*Stahlhelm*) in particular appear to have been singled out for lighter punishment.

Table 6.3 reveals three other types of resisters who, like communists, on average attracted harsher sentences from the People's Court: the 20 July group, members of

[14] From a sample of 1233 defendants, Schlüter (1995, p. 38) reports a higher figure for the overall share of death sentences, 33.6%. However, his figure is for *convicted* defendants, not *tried* defendants, which our figure of 26.6% relates to. Our corresponding figure taken from our sample of 4023 convicted defendants is 28.8%. We suspect that the remaining differences between the two figures for the share of death sentences are due to the very different sample sizes we each use.

Table 6.3 Trial outcomes by the main resistance groups

	Overall	*KPD/KPÖ*	Non-*KPD/KPÖ* communist	*SPD/SFÖ*	Other socialists	Schwarze Front	Bündische Jugend	German catholic groups	German national-conservative	20 July group	White Rose	Youth groups	Austrian CCL	Not organised
Execution														
January 1933 to June 1941	2.4	2.8	0.0	1.0	0.0	2.2	0.0	0.0	9.1	–	–	–	–	13.6
July 1941 to April 1945	45.5	50.7	31.1	31.6	35.0	–	0.0	20.0	11.1	56.6	28.0	63.0	20.1	54.1
Penitentiary														
January 1933 to June 1941	63.4	63.5	69.7	61.5	75.0	60.4	30.8	37.5	45.5	–	–	–	–	45.5
Mean length of penitentiary term (months)	81.7	85.1	76.6	74.4	68.9	67.8	84.0	72.0	124.8	–	–	–	–	93.0
July 1941 to April 1945	39.7	38.4	54.9	63.2	39.8	0.0	70.0	5.0	77.8	18.4	24.0	14.8	53.1	32.5
Mean length of penitentiary term (months)	88.5	96.2	92.3	74.0	90.0	0.0	78.0	60.0	65.1	61.7	80.0	72.0	59.6	82.0
Prison														
January 1933 to June 1941	24.5	24.2	24.6	26.4	20.5	23.1	38.5	25.0	45.5	–	–	–	–	31.8
Mean length of prison term (months)	19.9	19.7	19.1	20.0	20.3	21.6	23.4	21.0	18.6	–	–	–	–	20.6
July 1941 to April 1945	8.0	5.6	2.5	5.3	13.6	–	20.0	55.0	11.1	7.9	36.0	22.2	17.7	5.7
Mean length of prison term (months)	27.2	23.4	9.3	36.0	24.7	–	15.0	6.8	27.0	30.0	24.0	76.0	37.3	32.0
Dismissed or acquitted														
January 1933 to June 1941	9.7	9.5	5.7	11.1	4.5	14.3	30.8	37.5	0.0	–	–	–	–	9.1
July 1941 to April 1945	6.8	5.3	11.5	0.0	11.7	–	10.0	20.0	0.0	17.1	12.0	0.0	9.1	7.6

youth groups and resisters not formally attached to any organised resistance group. Indeed, these three varieties of resisters faced the highest rates of execution of all: 56.6%, 63.0% and 54.1%, respectively, in the period from July 1941. The brutal response of the regime to those who attempted to assassinate Adolf Hitler in July 1944, of course, is well known, and the reasons behind it understood. The high rate of death sentences handed out to youth groups and unorganised resisters, however, is perhaps less expected and somewhat harder to account for. One explanation may have been that the independent and unpredictable nature of such resistance (i.e. youth and unaligned), which could not be easily slotted into the normal political framework of the era, fundamentally challenged some of the regime's cherished conceptions of the national community (viz. that citizens who had not been politicised by the opposition and the nation's youth would be willing and proud members of it) and led it to cast such resisters as irredeemable malignancies that needed to be excised from the body politic post-haste.

In a similar manner, judges appear to have taken the seriousness of the threat posed by the act of resistance into consideration, distinguishing markedly in terms of punishment between major and minor forms of resistance. Compared to the somewhat less challenging actions of conspiracy to commit high treason or aiding and abetting, the more serious charges of assisting foreign powers, distributing illegal writings, crimes committed abroad, separatism and subversion of the armed forces brought with them a greater likelihood of execution or a longer term in a penitentiary. Moreover, those who engaged, or sought to engage in violence against the regime, and thereby posed a more immediate threat, were also more harshly punished than those whose resistance took a non-violent form.

People's Court judges, however, were not just sensitive to the level of political threat posed by the defendants before them. Indeed, one might say, in spite of the highly political nature of the trials over which they were presiding, judges also systematically took personal factors into consideration. Younger and older resisters, women, the better educated, those active in non-political and non-religious community organisations, married, or with no prior convictions for political crimes, or who showed 'good character' by cooperating with the authorities were all less likely to face the guillotine or long stints in a penitentiary. By the same token, judges, no doubt reflecting prevailing Nazi prejudices, tended to view those defendants with partial Jewish ancestry less favourably.

The case of juvenile resisters provides a good illustration of the nature and subtleties of judicial discretion in the People's Court (Geerling et al. 2013). While juveniles, ceteris paribus, received relative leniency and were on average about 20% less likely than others to face capital punishment, not all juveniles received such dispensation. Discernible patterns of judicial discretion in sentencing are evident even within the category of juveniles charged with high treason. Relative youth, a normal record in the Hitler Youth or the League of German Girls, the loss of one or both parents, no left-wing affiliations, the presence of adult co-defenders who may have 'misled' and having the good fortune of being tried under less punitive pre-war legislation usually brought juvenile defendants a modicum of mercy.

The severity of punishment meted out to juveniles increased with age. Though all defendants were not adults under German law at the time of their resistance, juvenile criminal law dictated that juveniles had to meet certain criteria—possession of intellectual and moral development comparable to that of an 18-year-old—before they could be treated as adults before the courts. Judges were at pains to demonstrate that their mentality was adult. Typically, that was easier to demonstrate in the older juveniles.

In this regard, precociousness could prove deadly. As Karl Engert, Vice President of the People's Court and presiding judge in the case of Helmuth Hübener, explained:

> Even the testing of his general knowledge, his extensive political knowledge, his competence as well as his appearance before the court and behaviour reflect the image of a precocious young man long grown out of youth. There is nothing to the contrary to suggest that his moral maturity had been retarded. Even in the sequence of events, there is no obvious sign that this was the act of a juvenile not fully matured. (Anklage 8 J 127/42 g, p. 19)

While not uncommon, such was not an automatic assertion of the People's Court. Relative youth could be a mitigating circumstance. In June 1944, the 17-year-old Austrian Johann Salzner was sentenced to a juvenile prison for an indefinite period (with a minimum of 2 years) for attempting to influence eight of his Hitler Youth comrades with his 'communist beliefs' while aged 15–16. By the standards of the People's Court, his sentence was comparatively light. He had been charged with listening to foreign broadcasts, spreading communist propaganda, attempting to undermine the fighting strength of the German army during war and attempting to change or alter the territory of the German Reich. The presiding judge, Johannes Heinrich Wilhelm Merten, explained his 'leniency' in terms of the defendant's relative youth:

> His behaviour fluctuated between extremes, as shown by his membership in a separatist organisation on the one hand and his inclination towards communism. It shows the need for self-assertion at this age and rebellion against the existing order born out of the preference for a type of Wild West romance. (Anklage 7 J 50/44, p. 3; Urteil 5H 37/44, pp. 3–4)

The fate of defendants, irrespective of their age, also seemingly depended on who was their presiding judge. Not all People's Court judges were the same in their sentencing behaviour. The depth of the ideological commitment of a judge to Nazism in particular appears to have been a factor influencing sentencing outcomes, especially in cases heard before 1942 (Geerling et al. 2017a). Judges, who had joined the party prior to its coming to power in 1933—and hence were more likely to have had ideological than opportunistic reasons for joining—were on average 3.6 percentage points more likely to impose a death sentence than other People's Court judges. Indeed, for each additional year of party membership, judges were 1.2 percentage points more likely to sentence a defendant to death.

Relative to their less ideologically committed brethren, these early members of the party, often called its old party fighters (*Alte Kämpfer*), were also 5.9 and 4.8 percentage points more likely to impose the death penalty on members of Comintern-aligned communist and social democratic parties, respectively. The targeting of these parties by *Alte Kämpfer* made political sense. As mentioned previously, from the regime's perspective, the greatest and most imminent threat for most of its existence lays in the two mass left-wing parties of the period, the *KPD* and *SPD*. These were parties that the *Alte Kämpfer* had literally fought on the streets during the years of the Weimar Republic and, in the 1930s and 1940s, they remained the political movements best positioned to challenge Nazi supremacy.[15]

Econometric analysis of sentencing data also unveils considerable evidence of other biases against defendants with specific characteristics (Geerling et al. 2017a). Compared to judges who did not join the Nazi Party before 1933, the *Alte Kämpfer* were more likely to sentence to death those who used violent means (17.4 percentage points), devout Roman Catholics (24.7 percentage points), defendants with partial Jewish ancestry (34.8 percentage points), juveniles (23.4 percentage points), the unemployed (4.9 percentage points) and foreigners (42.3 percentage points). The hatred of extremist Nazis towards Jews and foreigners more generally is, of course, well known. Yet, Nazi ideology was also deeply antagonistic towards all organised religion, especially those varieties, like Roman Catholicism, which sought to maintain their independence from Nazi control and influence. Individuals who attempted to assert Roman Catholic values in place of Nazi values were thus harshly persecuted. Furthermore, core to the Nazi worldview was the belief that those elements of German society which acted to weaken or hold it back from attaining its National Socialist destiny should be quickly and mercilessly expunged. As a consequence, one would expect that, unlike judges of other persuasions who might regard relative youth and economic disadvantage as mitigating factors at times of sentencing, the ideologically driven *Alte Kämpfer* would have shown markedly less sympathy or tolerance to juveniles or the unemployed who had chosen to challenge the regime's right to rule.[16]

There is also strong evidence that the sentencing decisions of People's Court judges could be swayed by events that were irrelevant to the case before them (Geerling et al. 2017b). Emotions can have a powerful and lasting influence on one's frame of mind and mood. Indeed, psychologists have demonstrated experimentally that strong emotions experienced in one domain can affect decision-making in another (Miller 2013, pp. 59–78; Schwarz and Clore 1983, pp. 513–523; Forgas and Locke 2005, pp. 1071–1081). Shock and anger, such as that generated when first learning of new heavy battle casualties or by contemplating the destruction of historic German cities in a major Allied bombing raid, could, therefore, have impacted the sentencing of individuals, who, while deemed to be public enemies, could not have had any direct responsibility for losses at the front or the destruction of cities.

Careful statistical analysis of sentencing outcomes confirms this possibility. The volume of *Wehrmacht* casualties in the month immediately prior to a sentence being

---

[15] Moreover, if the hometown of a judge who had joined the party before 1933 had lain in a centre of the Socialist Revolution of 1918 and 1919, he was ceteris paribus 5.8 percentage points more likely to sentence a defendant to death.

[16] For a good overview of Nazi ideology, see Kershaw (2015).

handed down in the People's Court influenced whether that defendant charged with treason or high treason would receive the death penalty. More precisely, it was found that a doubling in the monthly casualty figures on average coincided with a 5% increase in the likelihood of a death sentence being handed down. Such a deterioration in battle losses also decreased the chance of a defendant being acquitted. These findings held for battle fatalities up to 3 months before the sentence was handed down, with the effect most strongly felt 2 months after the increased casualty figures came out. Not all judges were equally likely to be influenced by battle deaths, of course. The effect was less apparent among judges who had previously served on the bench during the Weimar Republic, suggesting that their prior judicial experience had made them less susceptible to being influenced by bad news from the front. Certain defendants were particularly susceptible to this bias, too. Members of the *KPD* or *KPÖ* were more likely to see an increase in their chances of being sentenced to death than non-communists in the months following increased battle losses.

There is also evidence that the shock of particularly destructive air raids could have an effect on judicial deliberation long after the event had passed. The recent occurrence of one of six major Allied bombing raids on Germany—Berlin (25 August 1940), Cologne (30–31 May 1942), Bremen (25–26 June 1942), the Moehne dam (16–17 May 1943), Hamburg (24 July-2 August 1943) and Dresden (13–14 February 1945)—has been shown to have statistically impacted sentencing outcomes in the People's Court. A defendant unfortunate enough to be sentenced within a month after any of these events faced a 20% greater chance of receiving the death penalty.

Yet another factor that influenced sentencing was the defendant's behaviour in custody prior to appearing in court. Confronted by *Gestapo* interrogators determined to extract not only confessions of guilt but also information about their collaborators and contacts and what other activities had been planned, resisters faced a difficult choice: cooperate with the authorities and potentially receive a less severe punishment or remain quiet and feel the full wrath of the state. There was good reason to believe that the choice before them was real, too. Information obtained by the *Gestapo* through interrogations did benefit the defendants who provided it and was also used by the authorities against their co-defendants. Take, for instance, the case of two resisters from Hamburg, who after spending 14 months in custody, were both convicted in October 1938 of the high treasonous offences of belonging to a banned organisation and distributing illegal writings. Of these two resisters, one had cooperated and provided information to the authorities; the other had not. In sentencing, the defector received a 5-year penitentiary term with all 14 months of his time in custody deducted because, as the judge explained in summing up, he had 'confessed and helped clear up all the facts of the case'. By contrast, his co-defendant, who had chosen not to cooperate, was given an 8-year penitentiary term with only six of the 14 months spent in custody being counted against it. His harsher sentence was attributed by the judge to the fact that he had, despite 'the mountain of evidence, foolishly denied every activity', behaviour which only served 'to undermine his credibility'. The testimony of his co-accused, who had cooperated, however, was accepted by the court, 'as corresponding to the truth' (Urteil 1H 26/38, pp. 9, 15).

This was not an isolated incident. Statistical analysis of the outcomes of treason and high treason cases before the People's Court reveals that if a defendant cooperated with authorities, this usually brought them a degree of leniency at sentencing, reducing their term of incarceration by as much as 19.4% and increasing the prospect of their being sent to a normal state prison instead of a harsher penitentiary.

The temptation to cooperate with the authorities was thus strong. The responses of resisters to the personal dilemma that confronted them varied. Some, like Sophie Scholl, the 21-year-old student arrested when circulating anti-war pamphlets at the University of Munich in February 1943, revealed little in interrogation and even claimed sole responsibility for her actions in an attempt to protect other members of the White Rose, the non-violent resistance group to which she belonged (Scholl 1983; Sturms 2013). By contrast, many others provided the authorities with information. Erich Honecker, the leader of East Germany between 1971 and 1989, for example, is alleged, after being arrested for high treason in 1935 at the age of 23, to have provided incriminating evidence against fellow communists (Paterson 2011). If true, he would have been far from being the only resister in custody to have done so.

Table 6.4 reports the incidence of cooperation with the authorities across various characteristics commonly possessed by resisters. Overall, it reveals that 76.2% of resisters cooperated when in custody. Cooperation was slightly more common prior to the war and among female resisters. While the choice to cooperate or not was an individual one, Table 6.4 shows that teenagers, older adults, those charged with major high treason and treason, members of the Black Front (*Schwarze Front*) and enlisted soldiers were the most likely not to yield to the pressure. Yet, to understand better what attributes made some people more or less likely to cooperate requires deeper statistical analysis.

Such an analysis on a sample of resisters carried out by the authors elsewhere has uncovered a number of factors that indicate which individuals might have been more inclined to cooperate (Geerling et al. 2015). It found that cooperation was inversely related to the resister's highest level of educational attainment. For each 'step' in their attainment beyond a primary school education, the probability of cooperation decreased. A high school education made one 12.5% more likely to cooperate; a tertiary education a further 27.8%.[17] Two factors appear to lie behind this finding. Better educated resisters were, firstly, less inclined to capitulate to the intellectually bereft Nazi *Weltanschauung* and, secondly, less likely to be duped by *Gestapo* interrogation techniques.

[17] In other words, someone who had been at university was about 40.3% less likely to cooperate than some whose education did not extend beyond primary school.

Table 6.4 Rates of cooperation (share of defendants)

		Pre-war	Wartime total	1933–1945
	Overall	***82.9***	***72.1***	***76.2***
Gender	Male	82.1	71.1	75.4
	Female	90.2	78.4	82.0
Stage of life	Juvenile	66.7	72.2	71.4
	Young adult	85.0	83.6	84.4
	Mature adult	84.1	72.2	77.6
	Older adult	77.0	69.0	70.7
Charge	Major high treason	80.7	67.7	72.1
	Treason	64.3	62.4	62.5
	Minor high treason	86.0	86.1	86.0
	Previous terms of imprisonment	78.3	61.5	68.8
Resistance group and affiliations	*KPD/KPÖ*	82.5	68.9	75.5
	SPD/SPÖ	83.0	71.2	74.5
	German or Austrian national-conservative	100.0^	71.6	71.8
	Non-*KPD/KPÖ* communist	77.3	77.8	77.4
	Black Front	69.2	50.0	65.8
	Other socialists	83.3	68.3	76.5
	Motivated by strong religious convictions	76.9	72.3	73.2
	Non-political and non-religious organisation	84.3	71.5	76.8
	Trade union	83.4	69.6	74.5
Occupation	Student	93.8	81.0	83.8
	No particular occupation	87.9	78.4	80.8
	Blue-collar worker (skilled and unskilled)	83.6	72.2	76.9
	White-collar worker (excluding high-ranking state official)	82.5	76.2	78.6
	Professional (excluding military)	91.8	67.7	74.4
	Business or property owner	78.3	69.4	72.2
	High-ranked state official	50.0	82.2	80.9
	Engaged in full-time political work	79.7	63.9	70.5
	Enlisted soldier or non-commissioned military officer	0.0^	54.2	46.4
	Commissioned military officer	–	90.0	90.0
	Service worker	80.5	71.3	75.0
Highest educational achievement	Primary school	83.4	72.3	76.5
	High school medium	81.5	71.7	76.8
	High school advanced	78.2	75.3	76.8
	Tertiary	82.6	69.7	73.4
Vocational training	Completed or in the process	82.8	70.6	75.0

Note: The figures reported show the percentage of defendants in each category who, while in custody, cooperated with the authorities. A discussion on how cooperation can be determined from the official records can be found in Chapter 2. The mark '^' is used in the table to denote entries where there were very small numbers of individuals. Not too much should be read into those percentages

Deep commitment to a cause that was based upon a strong, often dogmatically interpreted, ideological, theological or philosophical value system fundamentally in conflict with the Nazis, such as Communism, Catholicism, Mormonism and Pacifism, was also a strong predictor of non-cooperation. Many with such convictions, of course, chose to defect anyway, but, from what we know about these movements and their belief structures, the idea that such a course of action represented a betrayal of the cause must surely have weighed heavily in the mind. Being an anti-war activist, for example, made one around 9% less likely to cooperate.

Political or religious causes were, of course, not the only affiliations that could impinge on the decision to cooperate. Skilled workers were among the most heavily unionised, organised and cohesive groups in twentieth-century Germany. Their training placed them apart from and above other sections of the workforce and encouraged a sense of camaraderie with fellow skilled workers.[18] It was a connection that saw many collaborate in acts of resistance against a regime which had abolished their unions and sought to constrain their independence and collective strength within the Nazi-run German Labour Front. Resisters who were skilled labourers were, as a consequence, about 15% less likely to defect.

Similarly, resisters who were members of voluntary non-political, non-religious organisations, such as the Red Cross, hiking societies, gymnastic clubs and sporting associations, exhibited a similar inclination towards cooperation. Often taken as measures of social capital, participation in these organisations, which proliferated in interwar Germany, also fostered trust between members and a sense of community, attributes which together appear to have engendered in some a belief that mutual silence was plausible (Satyanath et al. 2017; Putnam 2000; Durlauf and Fafchamps 2004). In fact, involvement in these organisations reduced the likelihood of cooperation by about 29%.

Another distinct group of resisters relatively resilient to the urge to cooperate were those aged between 15 and 17. This group had a strong sense of identity that overlapped, yet was separate from, those of their adult co-resisters. As discussed in Chapter 4, what distinguished juvenile from adult resistance were the degree of independent, small group decision-making involved; its atomistic nature; its frequent lack of strong adult supervision; and the absence of any attempt to fit actions into a broader strategy for change. Juveniles were also more idealistic, which appears to have made them more inclined to reject cooperation by as much as 25%.

While the idealism of youth may have created solidarity among juveniles, it appears not to have come at the expense of other attachments, least of all those of the family. Juveniles and young adults whose parents were both alive at the time of interrogation were about 20% more likely to cooperate. It was commonplace for *Gestapo* interrogators to threaten the well-being of family members. Unsurprisingly, teenagers and young adults, most of whom still lived at home and cared about their parents' welfare, were particularly vulnerable targets for such intimidation.

---

[18] It is noteworthy that levels of occupation (i.e. blue-collar, white-collar, professional, business and property owners) had no significant statistical impact on cooperation. It was the shared skills of artisans and their organisations that bound.

6.3 Conclusion

This chapter has provided glimpses of the resister's experience after being taken into custody. Some of the aspects it has addressed, such as the time taken to investigate and try cases, the resister's dilemma as to whether or not to cooperate with interrogators during this process and the determinants of the sentences they were ultimately to receive, have not, we believe, been previously considered in such detail by historians of resistance.

The chapter has also outlined the development of the legal framework and court system tasked by the regime with responsibility for dealing with treason and high treason. The cold, Social Darwinian calculus of national survival that lay at the heart of Nazism dictated that all parts of German society should ultimately come to reflect the political will and ideology of the Nazi state. For the judiciary, the practical meaning of this injunction was that its courtrooms should first of all be guided by the needs of the national community, not legal principle. Consistent with such a worldview, a new court, named the People's Court, came into being in April 1934, entrusted with the critical task of trying the regime's most serious political opponents. Yet, in the eyes of the party, this was not to be just another court, but an institution that would occupy a special place in the Nazi schema. As Karl Engert (1939, p. 485), one of the court's most prominent judges, explained: 'just as the *Wehrmacht* has to safeguard the external existence of the state, the People's Court has a similar obligation for inner security in collaboration with the *Gestapo*'. On his standing down as President of the Court in August 1942, Thierack expressed a similar sentiment to his successor 'in no other court does it emerge so clearly that the administration of the law…must be in accord with the leadership of the state. It will be your main task to guide the judges in this direction'. Consequently, People's Court judges should see themselves not just as jurists, but 'political fighters in the rebirth of the German nation' (Cited in Koch 1989, p. 127).

This chapter has used archival material and quantitative analyses to shine light on aspects of the reality of this highly politicised court. Were its judges in fact just political fighters for the regime, as Thierack had hoped, or did they (or some of them) retain, albeit buried beneath layers of National Socialist rhetoric, snippets of the judicial approach to the law? We have argued in this chapter that although the operation of the People's Court did reflect the arbitrary will of the state, its own court records indicate that it was far from indiscriminate in its processes. There was pattern and procedure for most of its history. More specifically, close analysis of the available statistical evidence suggests that in sentencing resisters, People's Court judges had considerable discretion, particularly at the margin. For the resister, that freedom could prove a double-edged sword, though. While it permitted judges to extend leniency, it also enabled them to deny it and, worse still, vent their anger at the course of the war by unfairly punishing those who stood before them in the dock.

Our findings thus lend weight to Karl Marxen (1994, pp. 34–87), Holger Schlüter (1995, pp. 140–85), Gruchmann (1988, pp. 956–68) and others whose work has advanced our understanding of the People's Court beyond crude notions of a blood tribunal where show trials were shamelessly played out. Although there is no doubt that from its inception the People's Court did dutifully carry out the will of the Nazi state, its decisions reveal that, at least up until mid-1942, its sentencing of defendants was neither indiscriminate nor a *complete* judicial farce. There still remained a semblance of legal procedure. Most importantly, the new quantitative and econometric evidence presented in this chapter suggests that even in sentencing political prisoners, individual People's Court judges, like judges in other systems, took a range of mitigating personal and contextual considerations into account. Even after controlling for the political backgrounds, resistance activities, criminal histories and personal characteristics of defendants, we have been able to detect statistically significant differences between the sentencing patterns of individual judges; some were more lenient, others less so.

When looking back at the history of the People's Court, thoughts of summary justice, execution and the total denigration of civil and legal rights rightly come to mind. It is difficult to disassociate the image of the ranting, merciless Freisler, who took over the presidency of the People's Court in August 1942, from the somewhat less theatrical and bloodthirsty origins of the court. That said, it cannot be denied that from its beginnings in 1934 as a provisional institution, the People's Court did represent the political will of the state. This does not mean, however, that the People's Court was a blood tribunal from the outset. As recent literature has shown, the conventional judicial and moral aspects of the legal system as well as legal integrity vanished under the presidency of Thierack, but in comparison with the late war years, the earlier history of the People's Court was still relatively moderate. Death sentences were rare, and certain vestiges of legal integrity remained, though, of course, at no time did the court exercise impartiality, fairness or justice. Yet, as the war progressed and further radicalised the Nazi regime, its rulers called on the nation's legal system to deliver ever more ruthless and expeditious forms of summary justice to their enemies. Fatefully, the People's Court and its judges, especially under the presidency of Freisler, were to prove themselves only too willing to adapt and oblige.

References

Primary Sources

(A) Archival Files

(i) Widerstand als Hochverrat

Anklage 5J 686/38 In: *Nationalsozialismus, Holocaust, Widerstand und Exil 1933–1945*. Online-Datenbank. De Gruyter. Accessed May 1, 2013. http://db.saur.de/DGO/basicFullCitationView.jsf?documentId=wh1116

Anklage 6J 166/41g In: *Nationalsozialismus, Holocaust, Widerstand und Exil 1933–1945*. Online-Datenbank. De Gruyter. Accessed Oct 7, 2013. http://db.saur.de/DGO/basicFullCitationView.jsf?documentId=wh596
Anklage 8J 127/42g In: *Nationalsozialismus, Holocaust, Widerstand und Exil 1933–1945*. Online-Datenbank. De Gruyter. Accessed Apr 27, 2013. http://db.saur.de/DGO/basicFullCitationView.jsf?documentId=wh2527
Anklage 7J 300/43 In: *Nationalsozialismus, Holocaust, Widerstand und Exil 1933–1945*. Online-Datenbank. De Gruyter. Accessed Feb 7, 2014. http://db.saur.de/DGO/basicFullCitationView.jsf?documentId=wh1177
Anklage 7J 302/43 In: *Nationalsozialismus, Holocaust, Widerstand und Exil 1933–1945*. Online-Datenbank. De Gruyter. Accessed Feb 12, 2014. http://db.saur.de/DGO/basicFullCitationView.jsf?documentId=wh2731
Anklage 7J 305/43 In: *Nationalsozialismus, Holocaust, Widerstand und Exil 1933–1945*. Online-Datenbank. De Gruyter. Accessed Feb 10, 2014. http://db.saur.de/DGO/basicFullCitationView.jsf?documentId=wh1847
Anklage 7J 50/44 In: *Nationalsozialismus, Holocaust, Widerstand und Exil 1933–1945*. Online-Datenbank. De Gruyter. Accessed Feb 12, 2014. http://db.saur.de/DGO/basicFullCitationView.jsf?documentId=wh1905
Anklage 11J 32/44 In: *Nationalsozialismus, Holocaust, Widerstand und Exil 1933–1945*. Online-Datenbank. De Gruyter. Accessed Oct 7, 2013. http://db.saur.de/DGO/basicFullCitationView.jsf?documentId=wh722
Urteil 1H 26/38 In: *Nationalsozialismus, Holocaust, Widerstand und Exil 1933–1945*. Online-Datenbank. De Gruyter. Accessed May 8, 2013. http://db.saur.de/DGO/basicFullCitationView.jsf?documentId=wh436
Urteil 1H 59/39 In: *Nationalsozialismus, Holocaust, Widerstand und Exil 1933–1945*. Online-Datenbank. De Gruyter. Accessed May 1, 2013. http://db.saur.de/DGO/basicFullCitationView.jsf?documentId=wh1117
Urteil 2H 137/41 In: *Nationalsozialismus, Holocaust, Widerstand und Exil 1933–1945*. Online-Datenbank. De Gruyter. Accessed Oct 7, 2013. http://db.saur.de/DGO/basicFullCitationView.jsf?documentId=wh597
Urteil 5H 104/43 In: *Nationalsozialismus, Holocaust, Widerstand und Exil 1933–1945*. Online-Datenbank. De Gruyter. Accessed Feb 12, 2014. http://db.saur.de/DGO/basicFullCitationView.jsf?documentId=wh2732
Urteil 5H 1/44 In: *Nationalsozialismus, Holocaust, Widerstand und Exil 1933–1945*. Online-Datenbank. De Gruyter. Accessed Feb 10, 2014. http://db.saur.de/DGO/basicFullCitationView.jsf?documentId=wh1848
Urteil 5H 37/44 In: *Nationalsozialismus, Holocaust, Widerstand und Exil 1933–1945*. Online-Datenbank. De Gruyter. Accessed Feb 12, 2014. http://db.saur.de/DGO/basicFullCitationView.jsf?documentId=wh1906
Zarusky, Jürgen and Hartmut Mehringer eds. 1998. *Widerstand als Hochverrat 1933–45. Die Verfahren gegen deutsche Reichsangehörige vor dem Reichsgericht, dem Volksgericht und dem Reichskriegsgericht*. Munich: K G Saur. Microfiche edition.

(B) German Laws

Reichsministerium des Innern. 1933a. Die Verordnung des Reichspräsidenten zum Schutz von Volk und Staat. *Reichsgesetzblatt* 28 February 1933, Part I, 83.
Reichsministerium des Innern. 1933b. Die Verordnung des Reichspräsidenten gegen Verrat am Deutschen Volke und hochverräterische Umtriebe. *Reichsgesetzblatt* 28 February 1933, Part I, 85.
Reichsministerium des Innern. 1933c. Das Gesetz gegen die Neubildung von Parteien. *Reichsgesetzblatt* 14 July 1933, Part I, 479.

Reichsministerium des Innern. 1934. Das Gesetz zur Änderung von Vorschriften des Srafrechts und des Strafverfahrens. *Reichsgesetzblatt* 24 April 1934, Part I, 341–48.
Reichsministerium des Innern. 1936a. Das Gesetz über den Volksgerichtshof und über die fünfundzwanzigste Änderung des Besoldungsgesetzes. *Reichsgestezblatt* 20 April 1936, Part I, 369.
Reichsministerium des Innern. 1936b. Das Gesetz zur Änderung des Strafgesetzbuches. *Reichsgesetzblatt* 2 July 1936, Part I, 1841.
Reichsministerium des Innern. 1936c. Die Verordnung gegen Wirtschaftssabotage. *Reichsgesetzblatt* 1 December 1936, Part I, 999.
Reichsministerium des Innern. 1938. Die Kriegssonderstrafrechtsverordnung. *Reichsgesetzblatt* 17 August 1938, Part I, 1455–57.
Reichsministerium des Innern. 1939a. Die Verordnung über außerordentliche Rundfunkmaßnahmen. *Reichsgesetzblatt* 1 September 1939, Part I, 1683.
Reichsministerium des Innern. 1939b. Die Verordnung gegen Volksschädlinge. *Reichsgesetzblatt* 5 September 1939, Part 1, 2378.
Reichsministerium des Innern. 1939c. Das Gesetz zur Änderung von Vorschriften des allgemeinen Strafverfahrens, des Wehrmachtsstrafverfahrens und des Strafgesetzbuches. *Reichsgesetzblatt* 16 September 1939, Part 1, 1841.
Reichsministerium des Innern. 1940. Die Verordnung über die Zuständigkeit der Strafgerichte, die Sondergerichte und sonstige strafverfahrensrechtliche Vorschriften. *Reichsgesetzblatt* 21 February 1940, Part 1, 405.
Reichsministerium des Innern. 1941. Die Verordnung über die erweiterte Zuständigkeit des Volksgerichtshofes. *Reichsgesetzblatt* 10 December 1941, Part 1, 776.
Reichsministerium des Innern. 1942. Die Verordnung zum Schutz der Rüstungswirtschaft. *Reichsgesetzblatt* 21 March 1942, Part 1, 165.

(C) German Law Periodicals

Engert, Karl. 1939. Die Stellung und Aufgaben des Volksgerichtshofes. *Deutsches Recht* 1: 485–86.
Richter, Hans. 1934. Das Gesetz zur Änderung von Vorschriften des Strafrechts und des Strafverfahrens vom 24. April 1934. *Deutsche Justiz* 18–35: 606.
Thierack, Otto Georg. 1936. Aufgaben und Tätigkeit des Volksgerichtshofes. *Zeitschrift der Akademie für Deutsches Recht* 19–20: 856.
Weiβ, Wilhelm. 1935. Der Volksgerichtshof des deutschen Reiches. *Deutsches Recht* 5: 518.

(D) Other

General Development of German Law during the Nazi Period (see Subsequent Proceedings), vol. 3, The Justice Case. 1951: Washington: United States Government Printing Office.

Secondary Sources

Boberach, Heinz.1975. *Richterbriefe: Dokumente zur Beeinflussung der deutschen Rechtsprechung 1942–1944*. Boppard am Rhein: Harald Boldt Verlag.
Broszat, Martin. 1959. Zur Perversion der Strafjustiz im Dritten Reich. *Vierteljahreshefte für Zeitgeschichte* 4: 390–443.
Buchheit, Gert. 1968. *Richter in roter Robe. Freisler Präsident des Volksgerichtshofes*. Munich: List.

Durlauf, Steven N., and Marcel Fafchamps. 2004. Social Capital. *NBER Working Paper* 10485.
Evans, Richard. 1996. *Rituals of Retribution: Capital Punishment in Germany, 1600–1987*. Oxford: Oxford University Press.
Forgas, Joseph P. and Judith Locke. 2005. Affective Influences on Causal Inferences: The Effects of Mood on Attributions for Positive and Negative Interpersonal Episodes. *Cognition and Emotion* 19: 1071–1081.
Geerling, Wayne, Gary B. Magee and Robert Brooks. 2013. Faces of Opposition: Juvenile Resistance, High Treason and the People's Court in Nazi Germany. *Journal of Interdisciplinary History* 44: 209–234.
Geerling, Wayne, Gary Magee and Robert Brooks. 2015. Cooperation, Defection and Resistance in Nazi Germany. *Explorations in Economic History* 58: 125–139.
Geerling, Wayne, Gary B. Magee, Vinod Mishra and Russell Smyth. 2017a (forthcoming). Hitler's Judges: Ideological Commitment and the Death Penalty in Nazi Germany. *The Economic Journal*.
Geerling, Wayne, Gary B. Magee, Paul A. Raschky and Russell Smyth. 2017b. *Legally Irrelevant Factors in Judicial Decision-Making: Battle Deaths and the Imposition of the Death Penalty in Nazi Germany* (February 27, 2017). Available at SSRN: https://ssrn.com/abstract=
Geerling, Wayne, Gary B. Magee and Russell Smyth. 2016. Sentencing, Judicial Discretion, and Political Prisoners in Pre-War Nazi Germany. *Journal of Interdisciplinary History* 46 (4): 517–542.
Gruchmann, Lothar. 1988. *Justiz im Dritten Reich 1933 bis 1940. Verwaltung, Anpassung und Ausschaltung in der Ära Gürtner*, 2 volumes. Munich: Oldenbourg.
Kershaw, Ian. 2015. *The Nazi Dictatorship: Problems and Perspectives of Interpretation*. London: Bloomsbury Revelations.
Koch, Hannsjoachim. 1989. *In the Name of the Volk: Political Justice in Hitler's Germany*. London: I.B. Tauris and Co Ltd.
Köpke, D. 2011. *Der Volksgerichtshof: Im Namen des Volkes?: Eine Analyse im Kontext der Neuordnung der Justiz im nationalsozialistischen Deutschland*. Hamburg: Diplomica Verlag.
Lauf, Edmund. 1994. *Der Volksgerichtshof und sein Beobachter: Bedingungen und Funktionen der Gerichtberichterstattung im Nationalsozialismus*. Opladen: Springer Fachmedien Wiesbaden GmbH.
Marxen, Klaus. 1994. *Das Volk und sein Gerichtshof: eine Studie zum Nationalsozialistische Volksgerichtshof*. Frankfurt am Main: Klostermann.
Marxen, Klaus and Holger Schlüter. 2004. *Terror und "Normalität": Urteile des nationalsozialistischen Volksgerichtshofs 1934–1945. Eine Dokumentation*. Duesseldorf: Justizministerium des Landes Nordrhein-Westfalen.
McDonough, Frank. 2015. *The Gestapo: The Myth and Reality of Hitler's Secret Police*. London: Coronet.
Miller, Michael K. 2013. For the Win! The Effect of Professional Sports Records on Mayoral Elections. *Science Quarterly* 94 (1): 59–78
Müller, Ingo. 1987. *Furchtbare Juristen. Die unbewältigte Vergangenheit unserer Justiz*. Munich: Kindler.
Ortner, Helmut. 2013. *Der Hinrichter: Roland Freisler – Mörder im Dienste Hitlers*. Frankfurt: Normen Verlag.
Paterson, Tony. 2011. Honecker was forced to resign by Secret Police. *The Independent*, June 6.
Picker, Henry. 1951. *Hitler's Tischgespräche im Führerhauptquartier 1941–1942*. Bonn: Athenseum-Verlag.
Putnam, Robert D. 2000. *Bowling Alone: The Collapse and Revival of American Community*. New York: Simon & Schuster.
Richter, Isabel. 2001. *Hochverratsprozesse als Herrschaftspraxis im Nationalsozialismus: Männer und Frauen vor dem Volkgerichtshof 1934–1939*. Muenster: Westfälisches Dampfboot Verlag.
Satyanath, Shanker, Nico Voigtländer and Hans-Joachim Voth. 2017. Bowling for Fascism: Social Capital and the Rise of the Nazi Party. *Journal of Political Economy* 125 (2): 478–526.

Schlüter, Holger. 1995. *Die Urteilspraxis des nationalsozialistischen Volksgerichtshof.* Berlin: Duncker & Humblot.
Scholl, Inge. 1983. *The White Rose: Munich, 1942–1943.* Trans Arthur R. Schultz. Middletown, CT: Wesleyan University Press.
Schwarz, Norbert and Gerald L. Clore. 1983. Mood, Misattributions, and Judgements of Well-Being: Informative and Directive Functions of Affective States. *Journal of Personality and Social Psychology* 45 (3): 513–523.
Steinweis, Alan E and Robert D. Rachlin. 2013. *The Law in Nazi Germany: Ideology, Opportunism and the Perversion of Justice.* eBook: Berghahn Books.
Sturms, Frank. 2013. *Die Weiße Rose: Das Schicksal der Geschwister Scholl.* Wiesbaden: Marix Verlag.
Wachsmann, Nikolaus. 2004. *Hitler's Prisons: Legal Terror in Nazi Germany.* New York: Yale University Press.
Wagner, Walter. 1974. *Der Volksgerichtshof im nationalsozialistischen Staat.* Stuttgart: Deutsche Verlags-Anstalt.
Weitz, Eric. 2009. *Weimar Germany: Promise and Tragedy.* New Jersey: Princeton University Press.
Wieland, Günther. 1989. *Das war der Volksgerichtshof. Ermittlungen-Fakten Dokumente.* Pfaffenweiler: Staatsverlag der Deutschen Demokratischen Republik.

Chapter 7
Impacts and Implications

In 1947, the German author, Hans Fallada, published what is now widely regarded as one of the classic novels about German resistance, "Everyone dies alone" (*Jeder stirbt für sich allein*). It tells the story of two Berliners, Otto and Anna Quangel, who, after learning of their son's death during the invasion of France, decide to strike back at the regime by leaving anonymous postcards and letters denouncing Nazi rule at various locations across the city. Hoping that these would be read, passed on and ultimately foment rebellion in the capital, all but a handful are in fact promptly handed into the police, and the couple are eventually apprehended by the *Gestapo*.

> What did you expect anyway, Quangel? You, an ordinary worker, taking on the *Führer*, who is backed by the Party, the *Wehrmacht*, the *SS*, the *SA*? The *Führer*, who has already conquered half the world and will overcome the last of our enemies in another year or two? It is ludicrous! You must have known you had no chance! It's a gnat against an elephant. I don't understand it, a sensible man like you!
>
> No, and you will never understand it, either. It doesn't matter if one man fights or ten thousand: if the one man sees he has no option but to fight, then he will fight, whether he has others on his side or not. I had to fight, and given the chance I would do it again. Only I would do it very differently.[1]

Seventy years after this fictional exchange between Inspector Escherich and Otto Quangel was penned, the mystery of, and quest to understand, those who risked all to resist remains not only fresh and alive but beckoning. Historical research of resistance, of course, has greatly expanded over the intervening seven decades, deepened and enrichened our understanding of the phenomenon, but like all healthy fields of research, mysteries and gaps not only persist but are constantly being opened up. This book has sought to add to that accumulated knowledge by attempting to tease

[1] Fallada (2009, pp. 376–77). The novel was in fact inspired by the real cases of Otto and Elise Hampel, a factory worker and domestic servant residing in the Berlin district of Wedding, who were arrested on 20 October 1942 and sentenced to death by the People's Court for illegal writings and subversion on 22 January 1943. On 8 April 1943, they were beheaded in Plötzensee Prison.

W. Geerling, G. Magee, *Quantifying Resistance*, Studies in Economic History,
DOI 10.1007/978-981-10-6008-3_7

further insights from the archives with quantitative analysis. Our efforts, we believe, have both cast light on a number of important existing debates and identified new areas of study that call for greater research. In the remainder of this chapter, we briefly recapitulate some of those findings, drawing out some of their broader implications and opportunities for future research, and then conclude with a discussion and assessment of the impact of serious resistance.

7.1 Implications

This volume has presented a wealth of information about resistance, resisters and the courts which tried them. In this section, we consider in more depth the significance of what we believe are the most important things uncovered by the book. The choice of what is emphasised below is, of course, subjective and inevitably reflects the authors' own interests. Others may well find different dimensions, not discussed here primarily for the sake of brevity, equally interesting.

In Chap. 3, a new periodisation of resistance was posited, an implication of which, we believe, was that a greater proportion of historical attention in future should be directed towards understanding and appreciating better the earlier stages of the regime's interaction with serious resistance. After all, nearly 43% of arrests for treason and high treason occurred before the outbreak of war, 73.8% before 1943. We have argued that the great waves of arrests in 1933 and 1935, in particular, were pivotal moments in both the establishment of Nazi power and the history of resistance to it. They were not just endnotes to the turmoil of the Weimar era. Nor is it accurate to describe, as is so often the case, resistance as being effectively stagnant or evaporating between January 1936 and December 1940. Within the Greater German Reich and its annexed and occupied territories, it was not. With just under a quarter of all arrests made in this period, at an average rate of around 17 per month, a far better characterisation for the period would be one of relative stability in the volume of serious resistance. Similarly, 1942, with 14.8% of all arrests, emerges as the year of peak resistance, larger in terms of the volume of arrests than both 1943 and 1944. Such figures alone should caution one from assuming that serious resistance only really began in the aftermath of the crushing defeat of the Sixth Army at Stalingrad. Indeed, seemingly as significant, if not more so, to the rise of serious resistance was the unexpected, regime-shocking successes of the opening phases of the first Soviet winter counter offensive and the casualties it inflicted, approximately a year earlier (5 December 1941 to 30 April 1942).

Moreover, we have found that the experience of serious resistance in terms of its timing, nature and volume varied significantly not only between Germany and Austria but between regions of these countries. There were distinct hotbeds of resistance, too, such as Leipzig, Hamburg, Dresden, Prague, Graz and Duesseldorf, while the prime locations for activity were Berlin and Vienna, two veritable capitals of resistance. Thus, for many purposes, it is more accurate to talk of multiple resister's experiences within the Third Reich rather than a single one common across all its parts. Generally speaking, resisters in Germany and Austria saw their efforts as distinct, and there was indeed little interaction, let alone coordination, between groups active in each place.

The experiences of resisters were also meaningfully differentiated by non-political and non-geographical considerations. As discussed in Chap. 4, the patterns and prevalence of resistance varied by one's age, ethnicity, gender and socioeconomic status. That said, serious resistance was overwhelmingly an activity of a male, ethnically German, blue-collar worker aged in his 30s or early 40s. Yet beyond this norm, various other shades of resistance could, and did, blossom and make their contribution to the struggle against Nazism. Each of these strands had their distinctive and distinguishing features. The efflorescence of high-profile, middle-class, military and professional resistance predominately by older males in the last year of the war is perhaps the best known. But there were others. Foreigners, stateless people and those of partial Jewish ancestry, though less visible, were in fact disproportionately more active in serious resistance. Likewise, juveniles and females offered their own distinctive varieties of resistance, which have generally been under-recognised in the literature because they tended to depart to some degree from the archetypal image of the resistance fighter. A key theme of this book has been that resistance groups, and a fortiori resisters, were more than just the politics to which they subscribed. Indeed, we contend that much has been learnt and even more can still be learnt about resistance by both studying more systematically the personal characteristics, histories and identities of different types of resisters and acknowledging that the experience of resistance *for the resister themselves* varied in ways in addition to their differing political affiliations.

From the perspective of the regime, of course, all serious resistance whatever its hue, motivation, or location was part of a single, general national struggle against what it perceived as the "enemies of the people" (*Volksfeinde*). While administratively much of this struggle was perforce pursued at district and regional levels by officials, investigators and jurists specialised in the local manifestations of resistance, to the regime and the organisation responsible for state security from September 1939, the Reich Main Security Office (*Reichssicherheitshauptamt*), the overall "battle" was a coordinated national one aimed at guaranteeing the integrity of a unified National Socialist Reich. It is, therefore, legitimate and meaningful, while acknowledging national, regional and individual differences within the Greater German Reich, to talk of a singular, internal supranational resistance struggle against the regime.

This book has also devoted considerable attention to what might be called the post-resistance stage of the resister's experience, namely, that period between their arrest and the implementation of their sentence, if convicted, or the verdict, if not. In particular, it has reported the results of quantitative analyses that have shone new light on, firstly, the life-determining choice made by all detainees in *Gestapo* holding cells and interrogation rooms – to cooperate with the authorities or not – and, secondly, the determinants of sentencing outcomes.

With regard to the former, in analysing a context that is about as close as one can get to a real-world example of what social scientists, economists and philosophers call a prisoner's dilemma, we found that despite the enormous incentives for an individual to cooperate – nothing less than the securing of one's life and liberty – somewhere between one-in-four and one-in-five detained resisters remained quiet and did not divulge information about themselves or their collaborators to the *Gestapo* or prosecutors. Exploring econometrically what made such defendants behave as they did, we discovered that higher education levels, a strong commitment to a more dogmatic variety of ideology or theology, shared identities between co-defendants, the idealism of youth and having a strong sense of community were all factors that discouraged cooperation while in custody. Beyond the fate of resisters, though, these findings suggest more generally that investing in social capital and encouraging deeper engagement with civil society may be two ways in which communities can mitigate some of the baleful ramifications of individualistic, self-interested behaviour.[2]

The analysis of sentencing outcomes carried out in Chap. 6 is significant in that it adds weight to those who have argued for a more nuanced understanding of the workings of the People's Court. Among other things, it has revealed previously unnoticed systematic variations in the decision-making behaviour of its judges. We have found that over the entire life of the Court, its judges consistently factored – each to different degrees, of course – specific aspects of the defendant's background and life experiences into the sentences they handed down. Younger and older resisters, women, the better educated, those active in non-political and non-religious community organisations, or married, or with no prior convictions, or who showed "good character" by cooperating with the authorities were all more likely to receive some extent of leniency.

Moreover, sentencing was also shown to have been influenced by the depth of the judge's ideological commitment to National Socialist ideology (which ceteris paribus encouraged harsher sentencing), the extent of their judicial experience (which could predispose them to relative leniency) and their susceptibility to the transference of negative emotional feelings from a disturbing external context to their courtroom. Major Allied bombing raids on German cities and bad news from the front in the form of mounting battle fatalities were examples of such extraneous factors, which could induce an effect to the determent of the defendant being sentenced.

One consequence of this judicial discretion, and everything that influenced it, was a persistent variation in sentencing outcomes between judges. Take the pre-war period as an example. The People's Court judge, Ernst Jenne, who sentenced more than 5% of all defendants in these years and who had been a member of the Nazi Party since 1 December 1930, was ceteris paribus more likely to hand down a prison sentence to a political prisoner (as opposed to hard labour in a penitentiary or death) than the People's Court longest serving President, Otto Thierack. Likewise, Wilhelm Bruner, the foundation Vice-President of the People's Court, gave penitentiary sentences that were on average 63% shorter. Thus, although at all times they did dutifully carry out the will of the Nazi state and did became ever harsher in their sentencing especially from mid-1942, not all People's Court judges were merely

[2] For more on the broader implications of this study, see Geerling et al. (2015, pp. 125–39).

mindlessly implementing prearranged sentences. Rather they retained some degree of discretion, even when their decisions did not always meet with political approval.[3]

This book has identified a range of issues and questions in need of further research. Some have already been mentioned in the foregoing discussions, but two fundamental questions remain which, we believe, warrant close attention. First, there is the issue of causation. In Chap. 3, it was noted that there were considerable differences in both the volume and intensity of resistance between regions in Germany and Austria. Non-parametric statistical analysis suggests that these regional differences are unlikely to be accounted for simply by either the size of the region's population or the number of supporters of the main opposition parties who resided there. Their explanation appears to be more complicated and varied. It is a pivotal question, since providing such an explanation would not only help us understand more rigorously why some regions were more inclined to resist than others but also offer a unique window on the determinants of the level of resistance exhibited more generally. To the best of our knowledge, no systematic quantitative attempt to address the question of causation has been undertaken to this point. It is our belief that this vital matter is unlikely to be adequately tackled by a macro-analysis of aggregated resistance in all of the Greater German Reich or, for that matter, in Germany or Austria alone. The experience of resistance was too diverse to lend itself meaningfully to such a high-level approach. Instead, the findings of this book suggest an alternative research strategy that, in our view, would have a better chance of success. That approach would involve the undertaking of a series of in-depth, empirical comparative analyses of a selection of regions with high and low levels of resistance activity or, if the requisite data were available, a single panel study including all or most regions. Such an approach, especially if complemented by detailed prosopographical analyses of large databases of resisters, may unlock further secrets of resistance and reveal in a robust manner factors, which may account for the varying incidence of resistance in different locations. Of course, implementing such a strategy would be a major research project in its own right.

The second outstanding issue relates to the question of impact, namely, what effect did serious resistance actually have? Should all acts and varieties of resistance be considered of the same degree of historical importance, and, if not, how can one rationally, critically and accurately evaluate the impact, or contribution, of different resistance groups? As discussed in Chap. 1, these are questions rarely explicitly confronted in the recent literature. In the next section, we put forward a framework which, we contend, not only enables such an objective analysis of impact to be undertaken but also provides a new perspective on how resistance should be defined.

[3]We know that the decisions of a number of People's Court trials did raise concerns in official circles. In the period between April 1935 and August 1936, for example, no fewer than 18 of the court's verdicts were openly criticised by the Reich Ministry of Justice (*Reichsjustizministerium*) for their mildness. Adolf Hitler, too, later complained that the decisions of the People's Court "had not initially corresponded to his desired tough standards"; see Gruchmann (1988, p. 965) and Sweet (1974, pp. 314–29).

While the focus of this section, like the book itself, is on serious "fundamental resistance" to the regime, the framework proposed could, with simple modifications, be adapted to the study of dissent, opposition and nonconformity.

7.2 Impacts

"First they ignore you, then they laugh at you, then they fight you, then you win" is an aphorism often ascribed to Mahatma Gandhi, which describes the different stages of a winning strategy of activism. While there is no record of Gandhi ever uttering this precise formulation of words, in *Freedom's Battle*, he did write: "Ridicule is like repression. Both give place to respect when they fail to produce the intended effect…In a civilised country when ridicule fails to kill a movement it begins to command respect".[4]

While the Gandhian philosophy of *satyagraha* (insistence on the truth or non-violent resistance) from which these quotations are drawn clearly had no practical relevance to the German resistance to the Nazi state, on a conceptual level, it is useful for its identification of a crucial, perhaps at times forgotten, feature of all forms of resistance, namely, that resistance does not occur in isolation. Resistance is a behaviour that is elicited and which seeks reaction. It is thus best understood as part of an interactive process between those who resist and those who would be resisted.

The notion that an interaction existed between the Nazi regime and resistance, of course, has deep roots in resistance historiography. Peter Hüttenberger's (1977, p. 26) view that resistance was a function of the extent of asymmetrical relations of rule, that is, the degree to which there is an attempt to impose a system of total societal domination, clearly reflects the interaction. The notion of *Resistenz* is also underpinned by a response to the regime's desire to intrude on aspects of people's daily existence. Indeed, as Heinz Boberach and Manfred Messerschimdt (1984) have pointed out, what in the end constituted "resistance" was often in fact determined by the state. In practice, given the "highly fragmented nature of the Nazi political system", that meant, as Hans Mommsen (1991, p. 161) noted, "ultimately it was primarily the *Gestapo* that decided what resistance was". As a result, partaking in normal, seemingly harmless activities, such as listening and dancing to swing music, could be transformed into acts of subversion and resistance by the disapproval of the state and its secret police.

We concur with these views, which help us understand conceptually the link between the regime and society, though, merely note that their focus lies on just one

[4] Gandhi (2013, pp. 197, 199). While the sentence may not be attributable to Gandhi, a variant of the quotation, however, can be traced back to an address in 1918 by the American trade unionist, Nicholas Klein: "And, my friends, in this story you have a history of this entire movement. First they ignore you. Then they ridicule you. And then they attack you and want to burn you. And then they build monuments to you. And that, is what is going to happen to the Amalgamated Clothing Workers of America"; see proceedings of the Third Biennial Convention of the Amalgamated Clothing Workers of America (1918, p. 53).

aspect of interaction: how the actions of the Nazi State drew forth and delineated the boundaries of resistance activities. The key point is that for resistance to occur, there must first be something to be resisted. It is the prior actions of the resisted which provide the basic motivation and rationale for the response from those who wish to resist. This is not a trivial point. Responses not initiated by genuine prior action of "those to be resisted" are aggression rather than resistance. The pretext that Nazi Germany invaded Poland in response to Polish attacks on the German radio station at Gleiwitz on 31 August 1939 or the Red Army Faction (*Rote Armee Fraktion*) campaign of assassinations and bombings from the 1970s to the mid-1990s was in response to the alleged terroristic actions of an increasingly authoritarian Federal Republic (*Bundesrepublik*) both represent examples of aggressive behaviour (unconvincingly) portrayed as resistance.

But, as Gandhi's words earlier suggest, resistance requires more than the actions of something to be resisted; it also requires that the activities that resisters make in response must have an impact on the resisted. That, after all, is the point of resistance: to make a difference.[5] If there is no impact, it is hard to claim objectively that resistance has occurred other than perhaps subjectively in the mind of the individual carrying out the activity. Thus, one may characterise the process of resistance as the following chain of events:

For "resistance" to occur, each link in the chain must occur: a motivating action (or actions) by those who would be resisted, an induced response by those who decide to resist and a reaction from the resisted. The last link in the chain is important. Intellectuals discussing in secret the shape of a post-Nazi Germany or Austria or an individual, unbeknownst to the authorities, entertaining anti-regime thoughts with their closest friends may be certainly displaying discontent with the system, but it cannot be regarded as resistance. An advantage of setting a desideratum for resistance that insists on at least a modicum of impact on those resisted is that, in so doing, one clearly permits a wide variety of activities, not just the most famous resisters, such as the 20 July group, the Kreisau Circle, the White Rose and the communist cells, but also many other groups and individuals whose actions caught the attention of the *Gestapo* in some way, to be considered.

A key question becomes how does one gauge impact, especially in the context of universal failure? As the record testifies, all German resistance to the Nazis ultimately proved fruitless: Germany's liberation from Nazi terror was not to come internally but rather awaited the advances of the conquering Allied armies. At the simplest level, the intensity of the reaction from the resisted or the degree to which they have a measurable impact on their ability to realise their goals provides possi-

[5] Michael Voges' claim that the working-class struggle in factories for better wages was resistance precisely because it threatened the regime's goals and required a political response makes this point; see Voges (1981, pp. 329–84).

bilities. Hansen (2014, p. 332), for example, has recently attributed an important role to the post-20 July "disobeyers" who, in the closing stages of the war, refused to follow Hitler's instructions to destroy German industry and infrastructure rather than allow it to fall into the hands of the Allies. He contends that they "played a great and largely unrecognised role in the recovery of Germany and, therefore, of Europe". Gandhi's words above, too, suggest that we can think in terms of partial success; in other words, there is a scale of impact to resistance. One, for example, presumably progresses beyond the stage of being ignored, only by proving ever more troublesome to, and hence impactful on, the authorities.

Towards the bottom end of that scale would lie activity that was just noticeable to, and worthy of recording by, the security services. Here, one might include the reported reluctance of some to fly the Nazi flag outside their house as regularly as others on special occasions or the tactic of always carrying a shopping bag in each hand in public, so that one need not have to give the Hitler salute (Stephenson 2001, p. 45). As severity of impact increases, *Gestapo* and police investigations begin, widen in scope, and charges heard before lower courts or protective custody eventually come into play. Yet, while they brought retribution, these acts of resistance – inter alia protests and complaints about specific issues and displays of disobedience, negativity and surliness – typically did not draw the ire of the regime. As Jill Stephenson (2016, p. 20) noted: "the response of the Nazi authorities to certain kinds of protest – including complaints made in a public place – clearly suggests that, while they regarded it as irksome, they did not regard it as a threat to their overall authority, even when it challenged a policy imposed by the regime, at local level, at least".

At the top of the impact scale, one would find the state's most serious charges: treason and high treason, acts which both sought to undermine the power and integrity of the state. Indictment on these charges was reserved in Nazi Germany, as indeed everywhere, for those whom the authorities perceived as their most severe political threats. Conviction, as we saw in Chap. 6, usually meant execution or long terms of imprisonment. The act of laying charges of treason and high treason against a resister, therefore, provides us with a neat indication, or signal, that the regime regarded the person, their group and the activities in question as being among the most troubling it faced. Put differently, these acts of resistance, either individually or collectively, were having such impact, real or potential, on the regime that it felt it appropriate to respond in turn with the most extreme legal tools available to it. Moreover, it follows from the previous point that the more regularly these tools were utilised against a group, the more persistent and pressing their impact on the regime is revealed to have been. By approaching impact in this manner, that is, in terms of the severity of the reaction of those being resisted, we contend that measures of impact that are consistent, not based on the prior moral and political predispositions of the assessor, and are potentially inclusive of resistance from any group can be constructed.[6] We further contend that the frequency with which charges of treason and high treason were laid against groups of resisters provides such an objective and accurate measure of the relative impact of the serious resistance they were waging against the regime.[7]

In Chap. 5, the patterns of treason and high treason by resistance group over time in Austria, Germany and outside the Reich were presented. Reconfiguring these data in terms of political groupings and broad political persuasions (or none as the case may be) permits a sense of the relative impact of different groups active within the Greater German Reich, as well as its occupied and annexed territories to be gauged. As Fig. 7.1, which breaks down all arrests for treason and high treason by resistance groups, shows, the three most active groups were communist in orientation. Indeed, more than two-thirds of those arrested between 1933 and 1945 were working for the Communist Party of Germany (*Kommunistische Partei Deutschlands* or *KPD*) or Communist Party of Austria (*Kommunistische Partei Österreichs* or *KPÖ*). If we add in other varieties of communist resisters, the percentage rises to over 72%. The only other group with 5% or more of the total arrests was the Social Democratic Party of Germany (*Sozialdemokratische Partei Deutschlands* or *SPD*). On the right side of politics, the most active groups were Austrian Catholic-Conservative-Legitimists, a classification used by historians, it should be noted, to link a number of small groups rather than a single entity per se, the Black Front (*Schwarze Front*), and the 20 July group. Together, these three nationalist, conservative and rightist groups accounted for 9.31% of arrests for treason and high treason. Most groups accounted for less than 1%, while just over 4% of those arrested were individuals unaffiliated with any organised resistance group.

Figure 7.2 now aggregates these arrests by broad political orientation. Across the entire reign of the Nazi Party, 82.8% of arrests for serious resistance came from the left of German and Austrian politics, 10.2% from the right, 2.9% were unaligned,

---

[6] In principle, this practice of gauging the impact of an action through the severity of the regime's reaction to it could also be extended to less serious forms of resistance as well as protest, dissent and nonconformity. To do so, all that would be required would be an observable and objective scale of the state's reaction to the activities in question. Thus, milder forms of nonconformity, such as dressing slightly unusually, might attract comments and criticism from local authorities but leave the individual largely untouched, whereas nonconformist behaviour, which from the regime's perspective was more disturbing, such as listening to and adopting a jazz or swing lifestyle, may invite a harsher response: police intervention and criminal charges. From the differing degrees of reaction experienced, one could then infer that it was the latter variety of nonconformity that was having the greater impact on the regime.

[7] We, of course, acknowledge that the cases of important, high-profile resisters, such as Ernst Thälmann, Claus von Stauffenberg and Dietrich Bonhoeffer, did not proceed through such normal court channels. However, we regard each of these cases as exceptions rather than examples of the norm for how serious resistance was dealt with by the Nazi state. Moreover, had Stauffenberg not been summarily executed on the morning of 21 July 1944 following his impromptu and unauthorised court martialling by Colonel General (*Generaloberst*) Friedrich Fromm, he, like most of his surviving co-conspirators, almost certainly would have been tried for high treason before the People's Court.

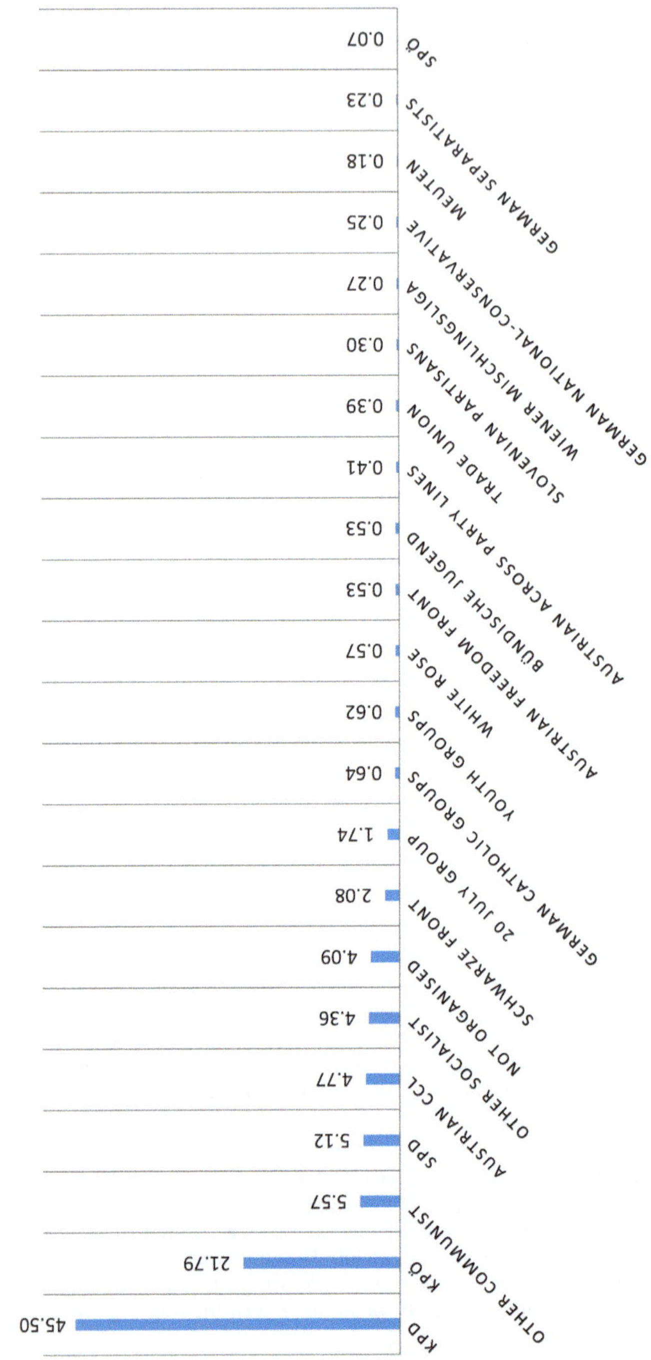

Fig. 7.1 Main resistance groups active in Germany, Austria and outside the Reich (percentage of total arrests), 1933–1945

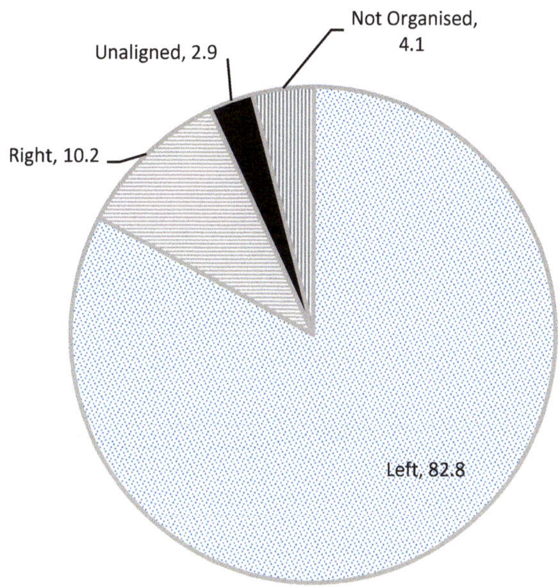

Fig. 7.2 The political orientation of resistance in Germany, Austria and outside the Reich (percentage of total), 1933–1945 (*Note*: The term "unaligned" refers to groups which were not formally affiliated with any of the main left- or right-wing parties, groups or organisations in Germany or Austria. They include those arrested for activities related to *Meuten*, the White Rose, youth groups, the *Wiener Mischlingsliga*, Austrian Across Party Lines groups, the Austrian Freedom Front and Slovenian partisans)

while the remaining 4.1% were unorganised. Figure 7.3 confirms that this overwhelming preponderance of serious resistance from the left was a feature in all years. In fact, the only period, in which the comparative contribution of the right to overall came within 30 percentage points of the left's, was in 1939–1940, when, in the aftermath of the *Anschluß*, there was a surge of Austrian Catholic-Conservative-Legitimist resistance.[8]

An examination of the political and group affiliations of civilians tried for treason and high treason in the Third Reich, thus, indicates that the vast majority of serious resistance *that elicited the most severe reaction from – and, hence, had the most impact on – the Nazi regime* originated on the left of German and Austrian politics. At least three-quarters, often more, of all resisters charged were from the

[8] The same quantitative dominance of serious resistance by the left played out in a broadly similar manner in each of the major settings of resistance. Thus, the relative shares of right and left resisters active in Germany, Austria and outside the Reich were 8.0 and 85.6 per cents, 14.7 and 77.1 per cents and 21.4 and 76.3 per cents, respectively. The shares of those arrested who were *KPD*/*KPÖ* members was 68.28% in Germany, 65.12% in Austria, and 51.58% outside the Reich. The only times and places where the left did *not* overwhelmingly dominate serious resistance were in Austria in 1939 and 1940 and outside the Reich in 1944. Not too much should be read into this last instance, though: there were only 24 arrests for activities outside the Reich in 1944 or, in other words, just 3.03% of all such outside arrests between 1933 and 1945.

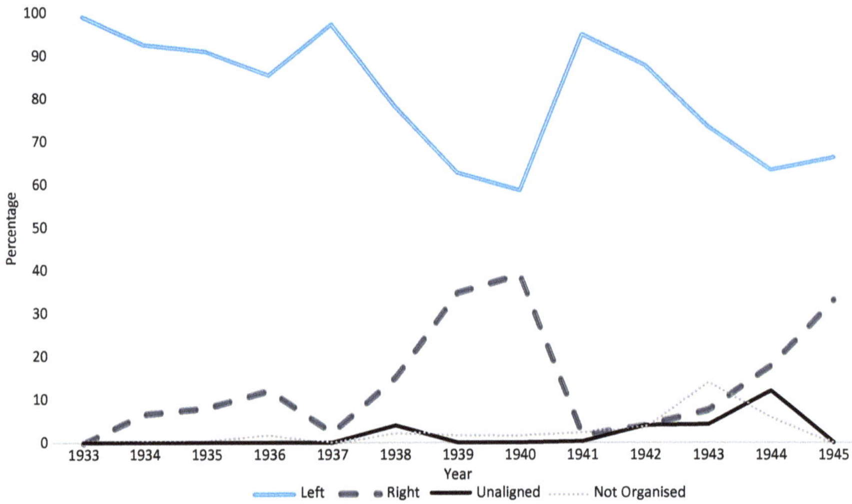

Fig. 7.3 Annual shares of arrests for treason and high treason by political orientation, 1933–1945 (*Note*: The term "unaligned" refers to the same groups from Fig. 7.2)

left, and in Germany proper (viz. Germany as it existed prior to the *Anschluß*), it was closer to 85%.

By contrast, resistance from the right, while making its presence felt, especially in 1939–1940 and 1944, figured considerably less prominently. It certainly had its high-profile resisters, more so than left resistance in fact, and could claim the most significant resistance events as its own, but quantitatively speaking, its contribution to overall resistance across the entire period of Nazi hegemony was not large.[9] To put that contribution further into perspective, it is worth noting that almost as many unaligned and unorganised resisters (7%) were arrested for treason and high treason as national conservatives in Austria and Germany combined (8.1%). Indeed, according to Fig. 7.1, unorganised resisters (4.1%) were the seventh most represented groups among the arrested, more common than every right-orientated group other than the Austrian Catholic-Conservative-Legitimists (4.8%). Looking at the facts presented in Figs. 7.1, 7.2 and 7.3, it is hard to escape the conclusion that across the entire period and in individual years, the groups whose serious resistance had the greatest impact on the regime, as evidenced by its reaction, were drawn from left-wing organisations and parties, among which the *KPD* and *KPÖ* were by far the most prevalent.

But what of the quality of resistance? After all, some acts or planned acts of resistance were more audacious in their goals than others. We will return to this

[9] Even at the peak of national-conservative resistance, this observation held true. As Wachsmann (2004, p. 222) noted, "Contrary to most people's presumptions, the men behind the 20 July plot on Hitler's life made up a minority among the Germans sentenced in the last year of the war. Communists, socialists and other members of the left-wing resistance were still much more likely to be sentenced by the People's Court than the national-conservative opponents of the Nazis".

issue below, but, for now, it is worth bearing in mind that it is hard to talk of a qualitative contribution to the impact of serious resistance when all of the more ambitious actions did not actually succeed. Nor should one confuse a particular (unsuccessful) event's chance of success with its impact. Indeed, if one is to talk of success, then it must be acknowledged that it was the more limited activities that did not pose an immediate existential threat to the regime – such as distributing anti-regime newsletters or trying to organise cells in the workplace – that at least for a while did succeed. It, of course, did not bring down the regime, but in a world of failure, small successes should not be casually dismissed.

Our analysis above and in Chap. 5 also confirms the frequently made observation that the left, in particular the *KPD*, was the first to experience the full wrath of the regime (Klönne, 2007, p. 144; McDonough, 2001, p. 5; Weitz, 1997, p. 6). On 20 February 1933, Hitler told a group of businessmen assembled at the villa of Hermann Göring that the time had come, whether by constitutional or other means, "to crush the other side completely".[10] The full means of the state was directed towards this task. The numbers of *KPD* and *SPD* members targeted were immense. At its peak in 1936, some 15,000 communist and socialists had been arrested (Merson 1985, p. 189). By 1939, as many as 150,000 of their membership had been interned in concentration camps (Kershaw 2015, p. 242). According to Wilhelm Pieck, by October 1935, 58% of the *KPD* cadre, the party's leadership, had been arrested, sentence or murdered by the regime. A further 30% had been forced to flee the country (Merson 1985, p. 182). Data on high treason and treason arrests heard before the People's Court reinforces this picture of the carnage experienced by the left in the pre-war years. Around 90% of those tried for these offences at that time came from left-wing parties, overwhelmingly the *KPD* and *SPD*.

Our analysis, however, provides a number of other findings that force us to rethink some assumptions about left-wing resistance at this time and beyond. First, it challenges common notions of the periodisation of communist resistance. In many ways, the assault on the left in the first 3 years of the regime in particular was a continuation of Weimar politics. But, as the data presented in this chapter and Chap. 5 illustrate, it would be mistaken to assume, as so often is the case, that the left, particularly the *KPD*, resistance in the pre-war years had effectively ended by 1935–1936 with its comprehensive destruction, the final, decisive act of the civil war between the right and left that had dominated Weimar politics. As Hansen (2014, p. 17) has argued "within two years [of the Nazis' coming to power], communist networks were riddled with informants, and, in 1935, a wave of arrests tore those resistance networks to shreds". In fact, *KPD* and left resistance of the most serious nature not only continued throughout the entire history of the regime, albeit at reduced rates between 1937 and 1940, but at all times represented in terms of volume by far its greatest threat.[11] On closer inspection, the periodisation of *KPD*

[10] Cited in Tooze (2006, pp. 99–100). See Benz and Pehle (1997, p. 59) for a similar quotation from the Reich Minister of War, Colonel Walther von Reichenau.

[11] Our findings, thus, provide a solid empirical and theoretical base for a view that has previously been expressed in parts of the literature; see McDonough (2001, p. 9) and Weitz (1997, p. 281), among others.

resistance, thus, appears more a situation of constant activity, albeit characterised by changing modes of resistance, than a matter of failed mass action in the first few years of the regime followed by relatively little activity thereafter.

Second, by drawing attention to the impact of resistance as gauged by the reaction it elicited from the regime, our findings raise doubts about the supposed inefficacy and wastefulness of communist resistance. As Benz (2014, p. 18) notes about communist propaganda tactics, “as risky and expensive in casualties as these actions were, their impact was small. While the National Socialist regime consolidated its power, the row of communists thinned out always quicker, without the copious amounts of circulated writings damaging the Nazis or bringing in the communist allies from other oppositional circles”. No doubt, this is true. Communist resistance did not seriously challenge the Nazis’ hold on power. Yet, it must be said that the same observation holds for all resistance. After all, groups such as the White Rose also engaged in pamphlet writing, which, while certainly heroic and brave, ultimately proved fruitless and led to its key participants’ executions. Even the 20 July group which came closest to assassinating Hitler failed in their attempt and in so doing not only led to the death and imprisonment of those associated with the plot but provided a short-term boost to Hitler’s popularity. Worse still, the failed coup engendered within the upper echelons of the *Wehrmacht* a commitment to undo the stain it had wrought to its character and honour, a development that no doubt made the final stages of war longer than they might have otherwise been.[12] The key point, it seems to us, is to view impact not from the perspective of whether or not it failed in its desired goal to remove the regime but rather the extent to which it caused the regime to confront, challenge and seek to destroy that resistance. Viewed from this perspective, it is hard to sustain the claim that communist resistance had less impact than resistance from any other group. Though never seriously challenging the security of the government, its sustained sabotage of the war effort, maintenance of illegal organisations, illegal leaflet distribution and financial support of the victims of Nazi terror did as much to undermine the totalitarian claims of the state as any other group or individual opposed to the National Socialism. Indeed, given the overwhelming preponderance of *KPD* and *KPÖ* members among those charged with high treason and treason across the entire duration of the regime, it might be reasonably concluded that their collective impact was all the greater.

Yet, to some, their efforts were pointless, for only a short, sharp and qualitatively different type of action such as the plot associated with Claus Schenk Graf von Stauffenberg could realistically offer the desired liberation from Hitler’s tyranny. As Kershaw (2009, p. 535) has succinctly explained, “Given Hitler’s outright dominance and unassailable position within the regime, significant change could by this time, autumn 1939, be brought about only through his deposition or assassination”. And if this were not to be carried out by those in his inner circle, then it could only be done by “a maverick individual operating entirely alone, outside any of the known – by now tiny, fragmented, and utterly powerless – left wing underground

[12] For a discussion of the effects of the failed bomb plot, see Kershaw (2012, pp. 30–35, 385, 389–90, and 394–6).

resistance groups which could so easily be infiltrated by the Gestapo" (Kershaw 2009, p. 537).[13]

We concur with much of the logic of this argument but wish to tease out some of the underlying assumptions upon which it is based. We have four points. First, by autumn 1939, Hitler had been in power for more than 6 years. If his power by that juncture had become "unassailable", then by extension, it had been not so, at least to the same extent, prior to that time and that, therefore, other means of ending the regime were possible (if still improbable) in that earlier period. Further, it encourages us to think more deeply and positively about the opportunities, perhaps missed, that existed in the regime's early years. Indeed, it is precisely then, in the immediate months and years after its inception, that it was at its most vulnerable. As we have shown, nearly half of all arrests for treason and high treason occurred in the pre-war era.

Second, it is important to remember that attempts on the life of Adolf Hitler and use of violent means more generally to bring about change were not restricted to the 20 July plotters and a few mavericks. Indeed, at least 52 planned or actual assassination attempts on Hitler are known of, most occurring towards the beginning of his rule (Moorhouse 2008, p. 2). Between March 1933 and the end of that year, for example, threats to kill Hitler, mostly from communists or social democrats, occurred on a weekly basis. And of these, the police regarded at least ten to have been serious. One such attempted assassin was Kurt Lutter, a communist carpenter, who planned with others to plant a bomb at one of Hitler's election rallies in March 1933. The Black Front, too, were involved in other similar plots (Weiss 1997, pp. 120–21). Of those convicted of treason and high treason, who used violent forms of resistance, 65.7% were from left-wing parties and groups (61.9% from the *KPD* and *KPÖ*).

Third, the belief that only an unaligned "maverick" could slip under the Nazi regime's security radar and benefit from some random stroke of good luck is highly questionable. Small, left-wing (and other) resistance groups were certainly under intense pressure and were constantly pursued and being infiltrated by the *Gestapo*. Such close scrutiny, however efficient, can never be infallible; nor is there any good reason for believing that small left-wing cells could not, like everyone else, experience random good fortune. The fact of the matter is that, then as in the present day, small independent cells of fighters can go undetected for long periods of time and through either a quirk of fate or glitch in the system find themselves presented with an opportunity to pose a major and unexpected threat to the state they oppose. If such a contingency had not actually been possible in Nazi Germany, the *Gestapo* presumably would not have placed so much effort and resources into ensuring that all hidden cells in their midst were identified and crushed. There was the lingering concern of the regime, too, that, while currently weak, at some point in the future

[13] Similarly, Anton Gill (2015, foreword) writes "…it was from within the army that the main Resistance to Hitler came, as a handful of determined officers perceived the evil towards which the Führer was leading the country".

popular support may well once again swing back to leftist parties. Better safe than sorry.

The fourth, and indeed the key, point we would like to make on this issue is that care should be taken not to conflate a particular conspiracy's chance of success with its broader place in the overall story of resistance. Indubitably, the 20 July plot was the single, most important event in the history of German resistance. It came closest to eliminating Hitler and removing the senior ranks of the Nazi leadership from power. Moreover, given that its key conspirators tried before the People's Court, when executed, were hanged with piano strings rather than being beheaded or shot – as was the norm for civilians and the military, respectively – it logically follows from the scale of regime reaction proposed earlier in the chapter that it was also the *resistance event* that had the greatest impact. But it should also be recalled that it was, after all, just one of many conspiracies in a sea of resistance that ebbed and flowed both before and after it. The 20 July plot should rightly be remembered and celebrated for the bravery of its participants, the brutal reaction it triggered from the regime and as a potent landmark in the struggle for a new Germany free from Hitlerism. It should not, even by default, be the template against which all other resistance should be assessed.

It is important to state clearly here that our argument is not the same one promoted by the state in the former German Democratic Republic as its foundational myth. We completely reject as factually incorrect the view that places the *KPD* at the head of a broad anti-fascist resistance movement and that leaves out or disparages the significant, independent contribution of non-communists. *KPD* tactics tended to hinder, rather than assist, the creation of a broad resistance front. Rather, our point is that, when looked at objectively through the visor of serious resistance cases heard before the People's Court, it is hard to deny the numerically dominant impact of the *KPD* on the Nazi regime across the entire period of its existence. The 20 July plot, of course, rattled the regime like no other event – and this showed up in the short, sharp spike in high treason cases related to it after July 1944 – but the deep concerns among the state's security apparatus engendered by *KPD* resistance persisted at all times, even in the last phase of the war. One can dispute whether or not such fears were justified; it cannot be denied the fears existed and the state acted upon them.

Our analysis, we believe, also calls for a fuller discussion of the manner in which overall accounts of German resistance are currently presented. While we embrace the inclusion of all forms of opposition in the resistance story, we do query whether uncritical inclusivity in itself provides a sufficient basis for balanced and nuanced understandings of resistance. It may in fact present a distorted picture. To take one highly visible example, in July 2014, a new permanent exhibition at the German Resistance Memorial Centre (*Gedenkstätte Deutscher Widerstand*) consisting of 18 rooms opened to the public. Of those rooms, just one was devoted to all resistance from the workers' movement combined and five to national-conservative activities. As the exhibition was housed in the Bendlerblock, the headquarters of the 20 July plot, its focus on that event is, of course, understandable. The image it projects through its implicit weighting of resistance activities is perhaps less so.[14]

For us, perhaps the most striking feature of our study of German and Austrian resistance has been the great diversity of people we have encountered who for one reason or another were drawn into it. If nothing else, this facet of the resistance story suggests that the attributes needed to stand up to the terror and brutality of the Nazi regime, whatever they may be, clearly did not belong solely to any one type of individual. Rather than being awestricken by the otherworldliness of the resister, we have instead come away from our research struck by their ordinariness, indeed by the very ordinariness of bravery itself. Most resisters, far from being the glorified and heroised figures often propagated, were in fact normal, everyday Germans and Austrians, with basic educations and ordinary blue-collar jobs, living standard lives, virtually unknown to all but their families, friends, neighbours and colleagues. What made them special was their involvement in the resistance.

Typical of the ordinary resisters who made up the vast majority of the serious resistance to Hitler's Germany was Josef Wager. Born in December 1905 to a working-class family from Augsburg, Wager's father had been badly wounded in the First World War and returned to Bavaria as an invalid unable to work. After completing a basic primary school education, Wager, who went by the nickname "Bebo", undertook and completed an apprenticeship as a turner at Firma Riedinger, joining the local branch of the metal workers trade union in 1923. In 1927, he took up full-time employment at the mechanical engineering company, the Augsburg-Nuremberg Machinery Factory (*Maschinenfabrik Augsburg-Nürnberg or MAN*), first as a fitter working on a lathe and then from 1929 as an electrician. In 1928, he married Lina Opalka and together they had two sons (Heinz and Helmut) and a daughter (Hanna) (Anklage 6 J 135/42 g, pp. 6–7; Urteil 6H 79/43, pp. 2–17).

Wager's political life began in 1923 at the tender age of 17, when he joined the Young Socialist Workers (*Sozialistische Arbeiterjugend* or *SAJ*) and the *SPD*. Within a short space of time, he had risen to the district leadership of the party's youth committee in Jakobervorstadt. Following Hitler's rise to power, he, along with friends from various *SPD* sporting, educational and youth associations from right across Southern Germany, decided to form a resistance group to be called the Revolutionary Socialists (*Revolutionäre Sozialisten*).

---

[14] Another example of such implicit weighting can be found in Benz (2014), a recent volume aimed at providing a broad overview of German resistance. The book is well-written and well-researched, yet, while acknowledging its requirement to cover all varieties of resistance, it is noteworthy – and far from unusual (e.g. see Nicosia and Stokes 2015) – that 39.17% of the book's content, measured in pages, is devoted to national-conservative resistance (Christian, traditional elites and military opposition) and 15.83% to left-wing resistance, of which the share of *KPD* resistance was just 5% (6 pages from 120). To be perfectly clear, we are not advocating here for a strict proportionality in the coverage of resistance groups in such books or exhibitions. Far from it, that would be cumbersome, intellectually constraining and unnecessary. Our point, rather, is merely that if the goal of a work is to provide an accurate representation of the history of resistance, then it should at least in some way acknowledge that proportionality. Not to do so may inadvertently mislead.

Established in the summer of 1933, the group's goal was simple: to overthrow Hitler. At first, the group's activities concentrated purely on smuggling forbidden *SPD* publications into Augsburg and its environs, believing that local support for Hitler could be undermined and ultimately turned against him by reading such material. The group also worked to create connections with other socialist organisations, most notably the New Beginning (*Neu Beginnen*).

In 1935, the Revolutionary Socialists (*Revolutionäre Sozialisten*), following the lead of the *SPD*, switched the focus of its efforts towards the provision of information about public opinion, developments and rearmament in Nazi Germany to *SPD* leaders then in exile in Prague. In 1937, the group once more changed tact, and together with like-minded comrades from Munich, it began planning armed attacks against the regime and the sabotage of armaments production. While the group continued to reject the notion of staging a putsch, it nevertheless felt it ought to be prepared for such a contingency, should propitious circumstances for such a course of action arise. By the summer of 1938, the group had acquired weapons and ammunition, which it began stockpiling in a bee house in Stätzling near Augsburg. With the advent of war, the group stepped up its sabotage campaign, targeting train lines and other key pieces of plant and infrastructure in Southern Germany and Austria. In the autumn of 1941, following the drafting of one of the group's other key co-founders, Hermann Frieb, into the *Wehrmacht*, Wager assumed overall leadership of the Revolutionary Socialists.

His time in charge was not to last long, though. Utilising information acquired from an infiltrated resistance group in Austria, which had had links with the Revolutionary Socialists, the *Gestapo* uncovered a stash of weapons hidden in Frieb's flat and arrested Wager while at work on 16 April 1942. He was 36 years old at the time. Charged with belonging to an illegal organisation, disseminating illegal writings, importing high treasonous materials, treason, and weapon offences, his trial before the People's Court began in Innsbruck on 22 March 1943. Although he had no prior criminal record, Wager was sentenced to death on 27 May 1943, 407 days after his arrest (the average processing time in this period was 377 days), and beheaded in Stadelheim Prison in Munich on 12 August 1943. His wife, Lina, who had been arrested at the same time, was released, as no evidence could be found to incriminate her. Wager's family, however, continued to be harassed by the *Gestapo* until the final days of the regime. In a farewell letter to his children, written on the day of his execution, Wager explained his actions in the following simple words: "I do not need to tell you I don't die as a criminal. I sought only good" (Nerdinger 1977, pp. 48–49).

It was the persistence of resisters, such as Bebo Wager and so many others like him, and their ongoing belief in the rightness of their cause, despite the personal risks and odds stacked up against them, that underscored the humanity of, and imparted the drive to, the resistance against National Socialism, ensuring more than anything else that its dystopian vision for Germany and Austria would not go unchallenged. In this most crucial of regards, those who resisted, despite their failure to dislodge Hitler, succeeded. It was a victory that lay not in the concrete achievements of a few but in the endurance of many.

References

Primary Sources

(A) Archive Files

(i) Widerstand als Hochverrat

Anklage 6J 135/42g In: *Nationalsozialismus, Holocaust, Widerstand und Exil 1933–1945*. Online-Datenbank. Accessed De Gruyter. Oct 16, 2013. http://db.saur.de/DGO/basicFullCitationView.jsf?documentId=wh2278

Urteil 6H 79/43 In: *Nationalsozialismus, Holocaust, Widerstand und Exil 1933–1945*. Online-Datenbank. De Gruyter. Accessed Oct 16, 2013.

Secondary Sources

Benz, Wolfgang. 2014. *Der deutsche Widerstand gegen Hitler*. Munich: C.H. Beck.

Benz, Wolfgang and Walter Pehle eds. 1997. *Encyclopaedia of German Resistance to the Nazi Movement*. New York: Continuum Publishing Company.

Boberach, Heinz ed. 1984. *Meldungen aus dem Reich*, volume 17, Berlin: Pawlak Verlag.

Fallada, Hans. 2009. *Every Man Dies Alone*. Trans. Michael Hoffmann. New York: Melville House Publishing.

Gandhi, Mahatma. 2013. *Freedom's Battle: Being a Comprehensive Collection of Writings and Speeches on the Present Situation*. The Floating Press (online).

Geerling, Wayne, Gary Magee and Robert Brooks. 2015. Cooperation, Defection and Resistance in Nazi Germany. *Explorations in Economic History* 58: 125–139.

Gill, Anton. 2015. *An Honourable Defeat: A History of German Resistance to Hitler, 1933–45* (eBook). Endeavour Press.

Gruchmann, Lothar. 1988. *Justiz im Dritten Reich 1933 bis 1940. Verwaltung, Anpassung und Ausschaltung in der Ära Gürtner*, 2 vols. Munich: Oldenbourg.

Hansen, Randall. 2014. *Disobeying Hitler: German Resistance in the Last Year of World War Two*. London: Faber and Faber.

Hüttenberger, Peter. 1977. Vorüberlegungen zum 'Widerstandsbegriff'. *Theorien in der Praxis des Historikers*, ed. Jürgen Kocka, 117–134. Göttingen: Vandenhoeck und Ruprecht.

Kershaw, Ian. 2009. *Hitler: A Biography*. London: Penguin.

Kershaw, Ian. 2012. *The End: Germany 1944–45*. London: Penguin.

Kershaw, Ian. 2015. *The Nazi Dictatorship: Problems and Perspectives of Interpretation*. London: Bloomsbury Revelations.

Klönne, Arno. 2007. *Jugend im Dritten Reich: Die Hitlerjugend und ihre Gegner*. Cologne: PapyRossa Verlag.

McDonough, Frank. 2001. *Opposition and Resistance in Nazi Germany*. Cambridge: Cambridge University Press.

Merson, Allan. 1985. *Communist Resistance in Nazi Germany*. London: Lawrence and Wishart Ltd.

Mommsen, Hans. 1991. The Political Legacy of the German Resistance: A Historiographical Critique. In *Contending with Hitler: Varieties of German Resistance in the Third Reich*, ed. David C. Large, 151–162. New York: Cambridge University Press.

Moorhouse, Roger. 2008. *Killing Hitler: The Plots, the Assassins, and the Dictator who cheated Death*. New York: Bantam.

Nerdinger, Eugen. 1977. *Flamme unter Asche. Die Gestapo schrieb die Geschichte der illegalen sozialdemokratischen Partei Südbayerns und Österreichs "Revolutionäre Sozialisten". Dokumente zu einer Lektion aus den Jahren 1933/45*. Augsburg: Stadt Augsburg.

Nicosia, Francis R. And Lawrence D. Stokes eds. 2015. *Germans against Nazism: Nonconformity, Opposition and Resistance in the Third Reich: Essays in Honour of Peter Hoffmann*. New York: Berghahn.

Proceedings of the Third Biennial Convention of the Amalgamated Clothing Workers of America (1918). New York: Cornell University Press.

Stephenson, Jill. 2001. *Women in Nazi Germany*. New York: Routledge.

Stephenson, Jill. 2016. Women and Protest in Wartime Nazi Germany. *Protests in Hitler's "National Community": Popular Unrest and the Nazi Response*, eds. Nathan Stoltzfus and Birgit Maier-Katkin, 18–37. New York: Berghahn.

Sweet, William. 1974. The Volksgerichtshof: 1934–1945. *Journal of Modern History* XLVI: 314–29.

Tooze, Adam. 2006. *The Wages of Destruction: The Making and Breaking of the Nazi Economy* London: Allen Lane.

Voges, Michael. 1981. Klassenkampf in der "Betriebsgemeinschaft". Die Deutschland-Berichte der Sopade (1934–1940) als Quelle zum Widerstand der Industriearbeiter im Dritten Reich. *Archiv für Sozialgeschichte* 21: 329–84.

Wachsmann, Nikolaus. 2004. *Hitler's Prisons: Legal Terror in Nazi Germany*. New York: Yale University Press.

Weiss, Herman. 1997. Assassination attempts against Hitler. In *Encyclopaedia of German Resistance to the Nazi Movement*, eds. Wolfgang Benz and Walter Pehle, 120–21. New York: Continuum Publishing Company.

Weitz, Eric D. 1997. *Creating German Communism, 1890–1990: From Popular Protests to Socialist State*. New Jersey: Princeton University Press.

Index

W. Geerling, G. Magee, *Quantifying Resistance*, Studies in Economic History,
DOI 10.1007/978-981-10-6008-3

MIX
Papier aus verantwortungsvollen Quellen
Paper from responsible sources
FSC® C105338

If you have any concerns about our products,
you can contact us on
ProductSafety@springernature.com

In case Publisher is established outside the EU,
the EU authorized representative is:
Springer Nature Customer Service Center GmbH
Europaplatz 3, 69115 Heidelberg, Germany

Printed by Libri Plureos GmbH
in Hamburg, Germany